The Calcitonins

Physiology and Pharmacology

M. AZRIA

THE CALCITONINS

Physiology and Pharmacology

124 figures, 66 tables, 1989

Basel · München · Paris · London · New York · New Delhi · Singapore · Tokyo · Sydney

Library of Congress Cataloging-in-Publication Data
Azria, Moïse, 1936-, The Calcitonins: Physiology and Pharmacology.
Bibliography: p.
1. Calcitonin. I. Title.
QP572.C3A97 1989 615'362 88-34813
ISBN 3-8055-4851-6

Drug Dosage
The author and the publishers have exerted every effort to ensure that drug selection and dosage set forth in this text are in accord with current recommendations and practice at the time of publication. However, in view of ongoing research, changes in government regulations, and the constant flow of information relating to drug therapy and drug reactions, the reader is urged to check the package insert for each drug for any change in indications and dosage and for added warnings and precautions. This is particularly important when the recommended agent is a new and/or infrequently employed drug.

© Copyright 1989 by
S. Karger AG, P.O. Box, CH-4009 Basel
Printed in Switzerland by Linsenmann AG, Basel

ISBN 3-8055-4851-6

Contents

VI

About the Author

Moïse Azria studied at the Research and Teaching Unit for Experimental and Human Biology of the Université René Descartes in Paris, graduating in 1961 and obtaining his doctorate in 1971. Since 1971 he has held regular teaching assignments at numerous academic institutions in France, including the Schools of Medicine and Pharmacy of the University of Paris, the Department of Pharmaceutical and Biological Sciences of the University of Nancy, the Institut Pasteur in Lille, and the Institut National des Sciences et Techniques Nucléaires at Saclay. He currently holds the title of Professor of Biology and Pharmacology.

Since 1975 Prof. Azria has been recognized as an *expert pharmacologue et toxicologue* by the French Ministry of Health, a capacity in which he is called upon to perform independent expertises on new drugs prior to their admission to the market.

In parallel with his activities in academia Prof. Azria has been involved since 1964 in many different fields of biological, toxicological, pharmacological and biopharmaceutical research and development in the pharmaceutical industry, where he is at present Deputy Head of Drug Delivery Systems with a leading multinational company. He has worked on calcitonin intensively for the past ten years and is currently engaged on a programme aimed at the development of non-injectable forms of administration.

Preface

When it was suggested that I should write a monograph on calcitonin it seemed "like a good idea at the time". I had after all been engaged in research involving calcitonin for more than ten years and was even then working on the development of non-injectable dosage forms of the hormone designed to overcome the principal obstacle to its greater acceptance as a major therapeutic agent. I felt I was bursting with knowledge about calcitonin and willingly accepted the challenge.

The formal task of putting it all down on paper, of course, proved more arduous than I had imagined, and took considerably longer. Lecturing at university level and working as a full-time research scientist in industry with its commercial and competitive pressures leaves little time for non-essential activities like writing books. This means that a considerable part of the burden of seeing this book finished and into print has been borne by others. They include fellow scientists, my laboratory and secretarial staff, and many others. It is in many ways invidious to single out individuals for special mention among all those who have contributed in greater or lesser degree, but I feel that I must acknowledge my particular indebtedness to three people without whose help and encouragement this book might never have been published – Joan Zanelli, Ph.D., for her technical advice and help, Marc Bleicher, M.D., for his part as my principal clinical consultant, and Stephen Cooper, B.A., for his work on the translation of my original French draft, for his tireless attention to detail, and for his part in the mammoth task of organizing and checking the references. I should also like to thank Lucien Chevrolet for converting my rough sketches into such clear schematic diagrams, and Roger Pittet and Hans B. Kälin for supervising the book's production.

My hope is that the result of all our labours will prove a useful, and easily digestible, source of information on the whole field of calcitonin and that it will challenge a new generation of researchers to solve its outstanding mysteries.

December 1988   Moïse Azria

Chapter 1: Introduction

Foreword
by Professor D. Harold Copp

Working late in the laboratory one evening in November 1960, I suddenly realized that we had evidence for a previously unrecognized hormone, released by high calcium perfusion of the thyroid-parathyroid apparatus of the dog, which lowered plasma calcium. I named it *Calcitonin* because it was obviously involved in regulating the level, or 'tone', of calcium in plasma. I was naturally excited by this discovery, but felt that it could not be particularly important (a view shared by my colleagues) or it would have been found much earlier. Dr. Azria's excellent monograph clearly indicates that I was much too cautious in my expectations. Calcitonin is now recog-

nized as a peptide of ancient lineage (it has been demonstrated in unicellular organisms such as *Escherichia coli* and *Candida albicans*) which has many important effects in addition to its anti-osteolytic action: it has proven therapeutic value in a number of disorders, including hypercalcaemia, Paget's disease and certain forms of osteoporosis. This monograph provides a comprehensive coverage of what has become a large and complex field. It is a major contribution to the endocrine literature.

D. Harold Copp

Since its discovery over 25 years ago, the hypocalcaemic hormone calcitonin has been extensively investigated in both animals and man. Findings have not always been consistent, however, and research may be expected to continue, leading possibly to the discovery of new therapeutic uses for the hormone and almost certainly to a fuller understanding of the physiology and pathology of bone – and perhaps of certain vascular diseases.

Calcitonin is an endogenous regulator of calcium homoeostasis, acting principally on bone. It also has a direct action on the kidneys and on gastrointestinal secretory activity, as well as direct and indirect effects on the central nervous system. In addition, investigations into its CNS role suggest that, besides an intrinsic analgesic effect, it may exert a modulator effect on neuronal activity, directly at the sites at which it is known to be present and indirectly, by a mechanism yet to be elucidated, at other locations.

Currently the principal indications for the therapeutic use of calcitonin are disorders involving hypercalcaemia, Paget's disease (osteitis deformans), acute pancreatitis, high-bone-turnover osteoporosis, pain associated with osteoporosis or bone metastases, and Sudeck's atrophy. Various types are in use – natural porcine calcitonin, synthetic human calcitonin, synthetic salmon calcitonin (Salcatonin) and a synthetic eel calcitonin analogue (Elcatonin) – one difficulty with treatment being that, until very recently, injection was the only possible mode of administration. Other dosage forms using other routes are now being developed, however, and a calcitonin nasal spray is in fact already commercially available.

Discovery: historical review

The first reference to the possible existence of such a substance as calcitonin was probably that made by Baber[1] in 1876. The second would have been made in 1925 if a communication submitted to the journal *Klinische Wochenschrift* had been published; as it was, the findings reported by Zondek and Ucko in that year did not appear in print until 1966[2]. Another early reference is the paper published by Nonidez[3] in 1932, but it was not until 30 years later, in 1961, that the existence of a second calcium-regulating factor in addition to parathyroid hormone was formally demonstrated, by Copp[4,5] in Canada. This discovery followed from the observation that when the thyroid and parathyroid glands of dogs were perfused with blood with a high calcium concentration, the concentration of calcium in the blood fell rapidly, indicating the release of some factor with a hypocalcaemic effect.

At the time Copp believed that this hypocalcaemic factor, which he called calcitonin, was secreted by the parathyroid glands[6]. The discovery of calcitonin was confirmed by MacIntyre's group in the UK in 1963[7], but in the following year Hirsch and his colleagues[8] in the USA reported that a hypocalcaemic substance was produced and secreted by the thyroid gland and not the parathyroids; consequently they named it thyrocalcitonin. Their conclusions were based on two observations in rats: firstly, that the fall in blood calcium was much greater after parathyroidectomy by cauterization than after excision by scalpel – probably because the former method also damaged the thyroid, releasing a hypocalcaemic factor in response – and, secondly, that injection of an extract of rat thyroid into another, young rat induced hypocalcaemia.

In the same year Foster and MacIntyre's team in London confirmed that the newly discovered hypocalcaemic factor was secreted by the thyroid gland and that Copp's calcitonin was identical with Hirsch and Munson's thyrocalcitonin by showing that perfusion of goat or dog thyroid with high-calcium blood was followed by a fall in blood calcium levels, whereas similar perfusion of the parathyroid had no effect[9]. They also found evidence[10], confirmed by Pearse[11] in 1966, that calcitonin – as it is now generally known – is produced and secreted by the parafollicular, or 'C', cells of the thyroid gland. Soon after, Pearse also showed that, in non-mammalian vertebrates, these cells are located in the ultimobranchial body, which has a common origin with the thyroid gland in the ultimobranchial cells of the primitive pharynx[12]. The three principal stages in the discovery of calcitonin are summarized in Figure 1.

This work was facilitated by the availability of a biological assay, developed by Hirsch and his colleagues[8], and greatly assisted by the interest shown by the pharmaceutical industry. This applies particularly

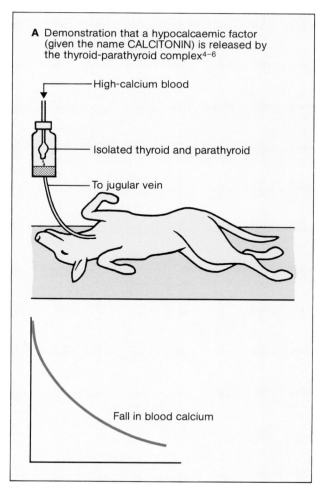

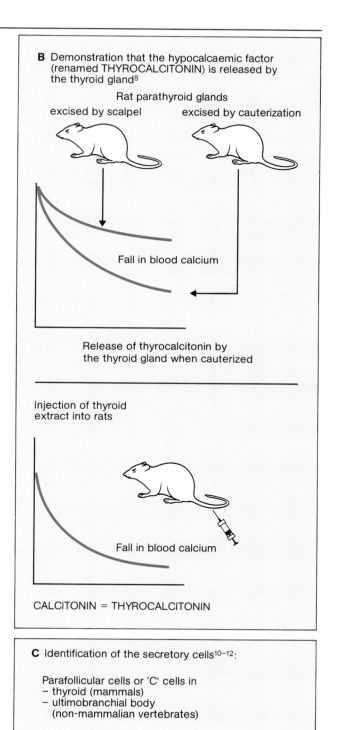

Fig. 1   Principal stages in the discovery of calcitonin

to salmon calcitonin after Copp's team, by arrangement with a local fish canning company (Canadian Fishing Co.), obtained over 100 kg of ultimobranchial tissue from about 500 000 salmon. From this material they were able to isolate sufficient calcitonin to determine its aminoacid composition[13], while its structure was elucidated by Potts' group at Boston[14] and the hormone was synthesized by Guttmann's team at Sandoz Ltd. in Basle[15]. This all took place within the space of 5 months. Meanwhile, MacIntyre's team had extracted human calcitonin from a medullary thyroid carcinoma (a C-cell tumour), and this quickly led to the synthesis of human calcitonin[16].

The different calcitonins

To date, calcitonins – or calcitonin-like substances – have been found in more than 15 species of mammals, birds, amphibians, fish and even unicellular organisms like *Escherichia coli*, *Candida albicans* and *Aspergillus fumigatus*[17], in which an immunoreactive human-like calcitonin has been detected using a combination of radioimmunoassay (RIA) and high-pressure liquid chromatography (HPLC). The calcitonins of eight species have been isolated in the pure state, while the structures of seven of these have been elucidated, and five of them have actually been synthesized. Many synthetic analogues have also been prepared in the search for a compound with improved therapeutic properties. Currently, however, only three calcitonins and one analogue[18] are in general medical use (Table 1).

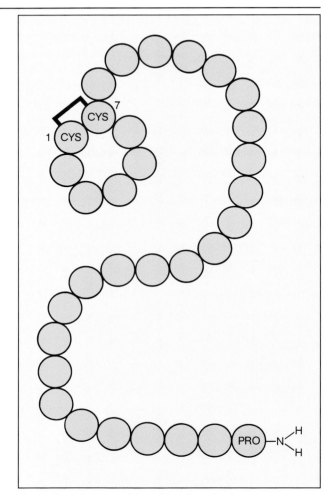

Fig. 2   Basic features of the calcitonin molecule

Species	Date of			Natural or synthetic substance
	Isolation	Structural elucidation	Synthesis	
Porcine (PCT)	1967	1968	1968	Natural
Human (HCT)	1967	1968	1968	Synthetic
Salmon (SCT)	1968	1969	1969	Synthetic
Asu1,7-eel (ECT)			1975	Synthetic analogue

Table 1   The principal calcitonins used therapeutically

The calcitonins are peptide hormones with molecular weights around 3500. The molecule is composed of a chain of 32 aminoacid residues with the following basic features (Fig.2):

– A disulphide bridge between the cysteine residues in positions 1 and 7, forming a ring of seven aminoacid residues at the N terminal, which carries a free amino group

– A proline amide group at the C terminal

These features (the number of aminoacid residues and the structure of the chain ends) are identical in all calcitonins and appear to be essential for biological activity. The central part of the chain, on the other hand, varies considerably from one calcitonin to another, both in the identity of the aminoacids present and in their positions in the chain, and it is these variations that are primarily responsible for the formation of specific antibodies.

The principal calcitonins may be divided into three groups according to their primary structure (Table 2), although similarity of structure is not necessarily reflected in similarity of activity[19]. Nor do structural similarity and origin appear to be particularly closely correlated. Salmon calcitonin, for example, has more aminoacids in common with human calcitonin than does the porcine variety, while human and rat calcitonins differ in only two aminoacids. One might have expected greater similarities within the group of mammals which are phylogenetically more closely related (Figs.3 and 4).

Group	Species	Number of aminoacids in common
I (artiodactyl)	Pig, ox, sheep	28
II	Man, rat	30
III (teleost)	Salmon, eel	29

Table 2   Calcitonins grouped according to structural similarity

Endogenous calcitonin occurs in a variety of forms, the precise biological significance of which is difficult to establish. For example, human calcitonin – like ACTH and the endorphins – may occur as an aminoacid sequence forming part of its inactive precursor, as a monomer or dimer in thyroid tumour cells[20,21], or in various immunoreactive forms in both healthy subjects and patients with medullary carcinoma of the thyroid[22].

Calcitonin gene expression

Recently, evidence has been found for tissue-specific RNA processing of calcitonin gene transcripts, with mRNAs encoding different peptide products, including precursors of calcitonin and calcitonin gene-related peptide ($_\alpha$CGRP and $_\beta$CGRP) respectively[23,24] (Fig. 5). It has also been shown[26–29] that human calcitonin is derived from a large precursor that also contains a calcitonin carboxyl-adjacent peptide (CCAP, also known as PDN-21 or katacalcin), which is useful as a tumour marker (Fig. 6). CCAP, which has no hypocalcaemic effect and no inhibitory action on bone resorption *in vitro*[29], is present in healthy subjects but is secreted in excess in medullary carcinoma of the thyroid. Indeed, it does not appear to have any physiological or pharmacological actions either *in vivo* or *in vitro*; all that is known for sure is that it is secreted along with calcitonin.

$_\alpha$CGRP, like $_\beta$CGRP[24,30], was predicted on the basis of complementary DNA sequence analysis and found to be a 37-aminoacid peptide with a 2–7 disulphide bridge and a phenylalanine amide at the carboxyl terminal[23–25] (Fig. 7). It is found mainly in the central nervous system[24,25,32], while calcitonin is most abundant in thyroid C cells. CGRP concentrations have been reported to be high in the spinal cord, amygdala and ventral striatum of the rat brain[25,33], and in some primary afferent fibres of the spinal cord CGRP is found at the same sites as substance P, suggesting a possible role in nociception[33–35]. In man it is distri-

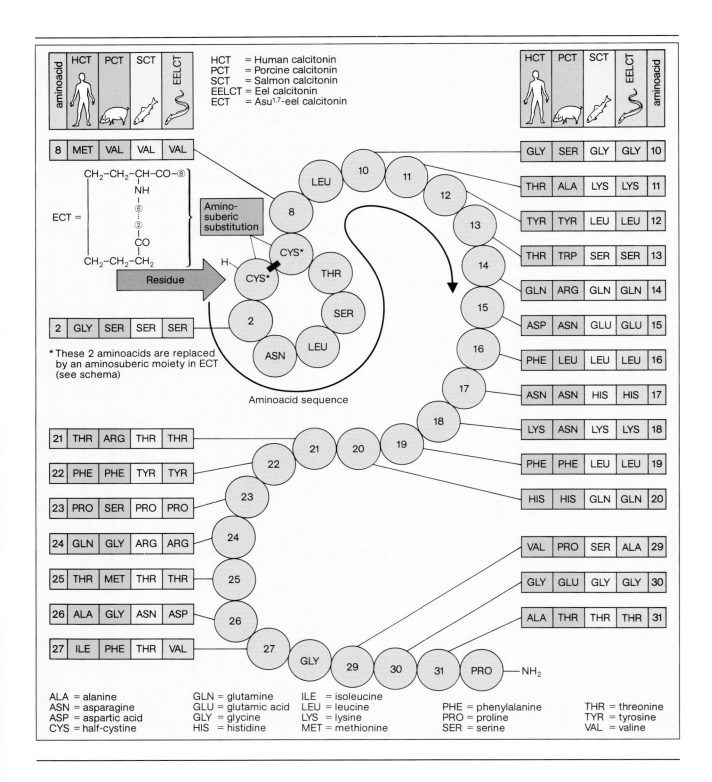

HCT   = Human calcitonin
PCT   = Porcine calcitonin
SCT   = Salmon calcitonin
EELCT = Eel calcitonin
ECT   = Asu^{1,7}-eel calcitonin

	HCT	PCT	SCT	EELCT	
8	MET	VAL	VAL	VAL	

$$ECT = \begin{array}{l} CH_2-CH_2-CH-CO-⑧ \\ \qquad\qquad\quad NH \\ \qquad\qquad\quad ⑥ \\ \qquad\qquad\quad ⋮ \\ \qquad\qquad\quad ② \\ \qquad\qquad\quad CO \\ CH_2-CH_2-CH_2 \end{array}$$

Amino-suberic substitution

Residue

	HCT	PCT	SCT	EELCT	
2	GLY	SER	SER	SER	

* These 2 aminoacids are replaced by an aminosuberic moiety in ECT (see schema)

Aminoacid sequence

	HCT	PCT	SCT	EELCT	
21	THR	ARG	THR	THR	
22	PHE	PHE	TYR	TYR	
23	PRO	SER	PRO	PRO	
24	GLN	GLY	ARG	ARG	
25	THR	MET	THR	THR	
26	ALA	GLY	ASN	ASP	
27	ILE	PHE	THR	VAL	

	HCT	PCT	SCT	EELCT	
	GLY	SER	GLY	GLY	10
	THR	ALA	LYS	LYS	11
	TYR	TYR	LEU	LEU	12
	THR	TRP	SER	SER	13
	GLN	ARG	GLN	GLN	14
	ASP	ASN	GLU	GLU	15
	PHE	LEU	LEU	LEU	16
	ASN	ASN	HIS	HIS	17
	LYS	ASN	LYS	LYS	18
	PHE	PHE	LEU	LEU	19
	HIS	HIS	GLN	GLN	20

	HCT	PCT	SCT	EELCT	
	VAL	PRO	SER	ALA	29
	GLY	GLU	GLY	GLY	30
	ALA	THR	THR	THR	31

—NH$_2$

ALA = alanine
ASN = asparagine
ASP = aspartic acid
CYS = half-cystine

GLN = glutamine
GLU = glutamic acid
GLY = glycine
HIS = histidine

ILE = isoleucine
LEU = leucine
LYS = lysine
MET = methionine

PHE = phenylalanine
PRO = proline
SER = serine

THR = threonine
TYR = tyrosine
VAL = valine

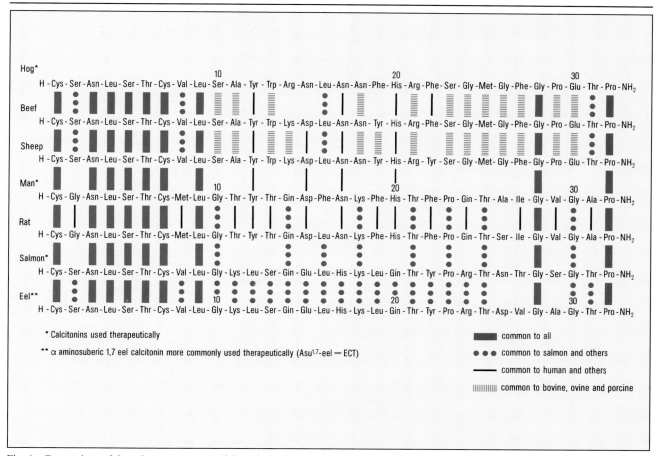

Fig. 4   Comparison of the primary structures of the principal calcitonins[19]

◀ Fig. 3   Comparison of the primary structures of human, porcine, salmon and eel* calcitonins

*The substance used therapeutically is Asu[1,7]-eel. This is obtained by replacing the 2 residues of half-cystine in the natural hormone by 1 residue of α-aminosuberic acid, thereby replacing the S-S linkage by an ethylene bridge, and by suppressing the N-terminal amino group[18].

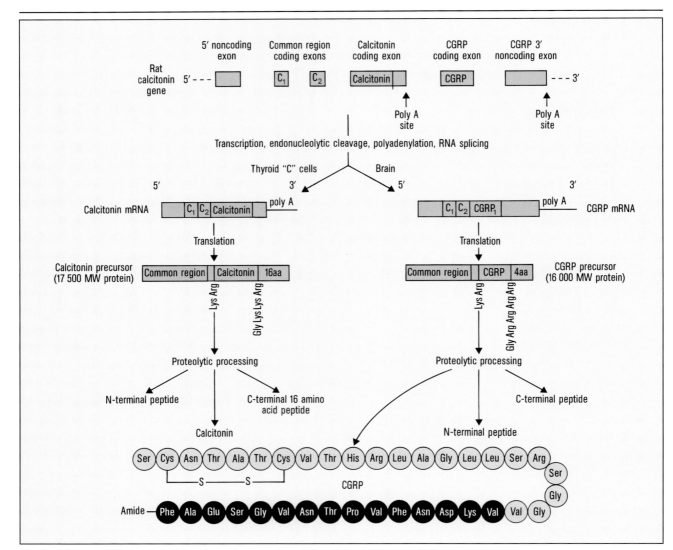

Fig. 5   Alternative RNA processing pathways in expression of the calcitonin gene, predicting the synthesis of a novel neuropeptide (CGRP) in the brain. The selection of alternative polyadenylation sites is suggested to account for the synthesis of calcitonin mRNA in thyroid C cells and the larger CGRP mRNA in the brain. CGRP mRNA from medullary thyroid carcinomas encodes a 16 000 MW primary translation product, which is probably processed to generate three polypeptides, including the 37-aminoacid polypeptide referred to as CGRP[25].

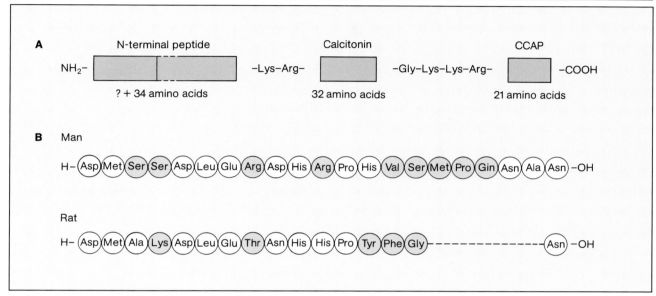

Fig. 6   A Position of calcitonin and calcitonin carboxyl-adjacent peptide (CCAP) within their common precursor
           B Aminoacid sequences of human and rat CCAP

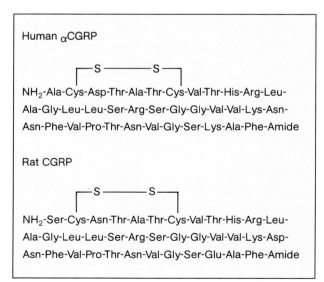

Fig. 7   Structures of human (α) and rat CGRP[31]

buted throughout the central nervous system (neocortex, cerebral cortex, periventricular mesencephalic region), with concentrations highest in the posterior grey matter columns of the spinal cord and in the pituitary[36-38]. High concentrations have also been reported[39] in the following areas of the human brain: the cerebellum, inferior olivary nuclear complex, certain parts of the central grey matter, the arcuate nuclei of the medulla oblongata and the dorsal motor nucleus of the vagus, with high densities of CGRP binding sites in the nucleus accumbens, amygdala, tail of the caudate nucleus, substantia nigra, the ventral tegmental area, the medial part of the inferior colliculus, the medial pontine nuclei, locus coeruleus, inferior vestibular nucleus, substantia gelatinosa of the spinal trigeminal nucleus, the nucleus of the solitary tract and the lateral cuneate nucleus. CGRP is also present in high concentrations in perivascular nerves throughout the body, including in the coronary[35] and cerebral vessels[40]. It also appears to be present in normal human thyroid[36,41] and abundant in medullary thyroid carcinomas[36].

The main properties of CGRP known at present are summarized in Tables 3 and 4 and in Figures 8 and 9. Its distribution in the central nervous system suggests a role as a neurotransmitter or neuromodulator and it has been reported to act on both the skeletal and cardiovascular systems[30,31,40,42,43]. In fact, the most striking effects reported for CGRP are cardiovascular, viz. an increase in the force and rate of contraction of isolated rat atrium[31,42] and extremely potent vasodilation in rabbit, hamster and man[31,44]. In healthy volunteers[45] CGRP by infusion causes hypotension and reflex tachycardia at plasma concentrations as low as 56 pmol/l, suggesting that it is more potent than the other known vasodilators. Infusion of CGRP into the coronary arteries (unpublished results by Maseri et al. cited in [39]) also results in pronounced vasodilation.

CGRP is the principal circulating product of the human calcitonin gene, strongly suggesting that it has an important physiological role in the control of blood flow and vascular tone, a control which may extend to the cerebral and coronary circulations[40]. Plasma levels are usually elevated in medullary thyroid carcinoma[41,46] and it is quite possible that the episodes of flushing[47] which occur in this disease are due to this peptide. Whether CGRP affects vascular tone

	Route of administration	SCT	HCT	CGRPs
Inhibition of				
Bone resorption	i.v/*in vitro*	+++	++	+
Gastric-acid secretion	i.c.v./i.v.	++	+	+
Food intake	i.c.v.	+	+	+
Pain perception	i.c.v.	++	+	+
Hypertension	i.c.v.	++	N.D.	++
Tachycardia	i.c.v.	N.D.	N.D.	++
Hypotension	i.v.	o	o	++
Myocardial contractility	i.v./*in vitro*	+/o	+/o	++
Vasodilatation	i.v.	+	+	+++
Renal adenylate cyclase activation	*in vitro*	++	+	+
Plasminogen activation	*in vitro*	++	+	+

Table 3   Main biochemical and pharmacological properties in man and rat of salmon and human calcitonins and of CGRPs (based on J. Fischer, personal communication)

N.D. = not done

Effect	Tissue/site/route of administration	Species	Mode of action
Positive myotropic effect	Isolated atrium	Rat	Peripheral
Relaxation	Aorta strips	Dog	Peripheral
Dilatation of microvessels	Cheek pouch	Hamster	Peripheral
Intense flushing	Intradermal	Man, rabbit	Peripheral
Tachycardia and hypotension	Intravenous	Man	Peripheral
Hypertension, tachycardia and rise in catecholamine production	Intracerebro-ventricular	Rat	Central

Table 4   Cardiovascular effects of CGRP (I. MacIntyre, personal communication, and author's unpublished findings)

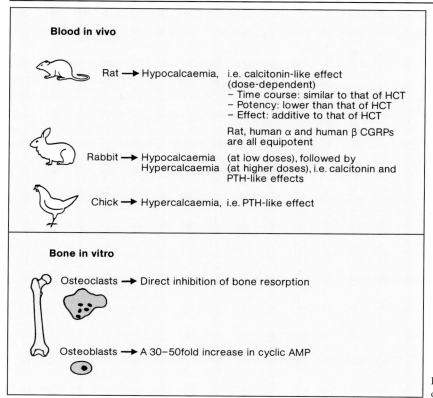

Fig. 8   Effects of αCGRP on blood levels of calcium and on bone

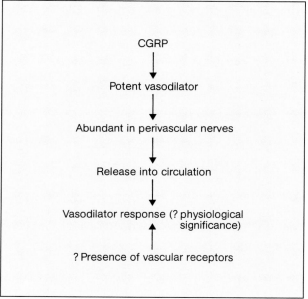

Fig. 9   Possible mechanism of action of CGRP on the cardiovascular system (I. MacIntyre, personal communication, and author's own [unpublished] findings)

in other diseases remains to be discovered, though the role of calcium as a co-antihypertensive (putative) substance might be due to an increase in CGRP secretion.

CGRP is known to have other effects similar to those of calcitonin but at much higher doses. It inhibits gastric secretion when administered directly into the central nervous system[48], and pentagastrin-, histamine- and bethanechol-stimulated gastric secretion when given intravenously[49]. Given intravenously to mice, human CGRP causes accumulation of cyclic AMP in the kidney and calvaria when measured using a microwave tissue fixation method[50]. These effects are dose-dependent. The time courses in the two tissues are different in terms of response to HCT and human 1–34 PTH (P. Bevis and J. Zanelli, personal communication).

Methods of calcitonin assay

There are at present basically two techniques for assaying calcitonin (Table 5):

Radioimmunoassay

This method can be applied either directly to biological fluids or after extraction from tissue or fluid. It measures immunoreactive calcitonin but not necessarily biologically active calcitonin (Fig. 10). Because of its sensitivity this is the technique most commonly used at the present time, although the accuracy of the results depends on the specificity of the antiserum used (no cross-reaction with other calcitonins, inactive forms of the hormone or with other hormones; see Tables 6 and 7), on the limit of detection (detection limits of 2 pg/tube for direct assay and of 0.2 pg/tube for extraction assay have recently been reported for endogenous HCT[52]; Table 8), and on the presence/absence of analytical artefacts (e.g. plasma matrix effect) and/or interference due to pathological conditions[53–55] – what Deftos has called 'an artifactual elevation of plasma calcitonin'[56].

Radioimmunoassay is an excellent technique for monitoring blood levels of the hormone and is commonly used to diagnose certain postmenopausal hypocalcitoninaemic states and, more especially, hypercalcitoninaemia associated with medullary carcinoma of the thyroid. It is also used in thyroid function tests to measure calcitonin secretion in response to calcium supplementation, to assess body reserves of the hormone, and to monitor the response to treatment. Finally, it also finds application in bioavailability and bioequivalence studies with different dosage forms of calcitonin as a therapeutic agent. More sophisticated RIA techniques are currently being developed for calcitonin, such as immunoradiometric (labelled antibody) one or two-site assays.

Biological assay

This is an indirect method of determination because it is based on measurement of the hypocalcaemic response ($\Delta\% Ca^{2+}$, normally after 60 minutes) to a parenteral (i.v. or s.c.) dose of calcitonin. It is performed in young rats under standardized experimental conditions (Fig.11) and, although older and less sensitive than radioimmunoassay, more accurately reflects true biological activity, or potency. Bioassay cannot be used to distinguish between different types of calcitonin, but proved invaluable during work on the purification of the hormone. It is still the only assay method approved by health authorities and given in pharmacopoeias (e.g. European Pharmacopoeia) for calcitonin in both the pure state (active ingredient) and as a constituent of dosage forms.

The effect of calcitonin on blood levels of calcium (total or ionized) is also used in clinical chemistry in the acute hypocalcaemia test[57], which is used to predict whether calcitonin is likely to be effective in certain bone diseases and thus to distinguish between

Principle	Method	Special features	Applications
1. Direct method Assay of CT in biological fluids	Radioimmunoassay (RIA) = assay of immunoreactive CT	Accuracy depends on: – Specificity of the antiserum – Analytical artefacts – Interference due to pathological conditions Does not distinguish between bioactive and non-bioactive forms of calcitonin and may not distinguish between different molecular weight forms	Clinical chemistry Assay of <u>endogenous</u> (basal level and fluctuations) and <u>exogenous</u> CT in biological fluids (plasma, CSF, urine) in: – <u>Diagnosis</u> (medullary thyroid cancer, hypo- and hypercalcitoninaemia), the CT acting as a marker – <u>Functional tests</u> (secretion, reserve): CT profile after infusion of Ca^{2+} – <u>Monitoring of effects</u> <u>of treatment</u> <u>Pharmacokinetics and</u> <u>biopharmaceutics</u> (bioequivalence studies)
2. Indirect method Evaluation of the potency of a calcitonin by measuring its principal effect, i.e. <u>hypocalcaemia</u> (determination of fall in blood level of total or ionized calcium after parenteral administration of CT)	Biological assay ('potency')	Less sensitive than radioimmunoassay Better reflection of 'endogenous' calcitonin status and activity Does not distinguish between different calcitonins or active fragments	Clinical chemistry – Induced hypocalcaemia test (blood calcium profile after injection of CT) – Indicates likely responders and non-responders to CT <u>Biopharmaceutics</u> – Standardization of the activity of dosage forms (expressed in IU)
3. Other methods Measurement of a biochemical variable other than calcium Radioreceptor assays Quantitative cytochemistry	Determination of (plasma, renal, urinary) cAMP, (plasma) phosphate	Technically difficult and time-consuming	Primarily for research purposes

Table 5   Assay methods for calcitonin

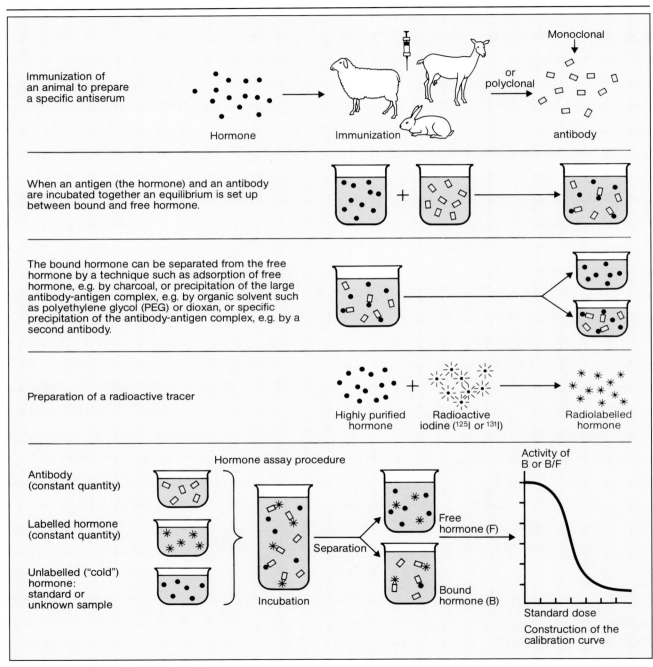

Fig. 10   Methodology of (hormone) radioimmunoassay

Antibodies	Cross-reaction with CT of types
Anti-porcine CT	Porcine, sheep, bovine and canine
Anti-human CT	Human, rat
Anti-salmon CT	Salmon I, II, III, rainbow trout, tuna, cod, Japanese eel, domestic fowl, pigeon, turkey, Japanese quail, green frog

Table 6   Cross-reactivity with antibodies to porcine, human and salmon calcitonin[51]

No cross-reactivity with doses up to 10 μg of any of these peptides			
Human	PTH	Porcine	Glucagon
	Insulin		Bradykinin
	Gastrin 1–17		Calcitonin
	Pentagastrin		VIP
	LH		GIP
	FSH		Motilin
	Prolactin		CCK 1–33
	TSH		
	Growth hormone	Bovine	PTH
	LHRH		Angiotensins I and II
	PP		Neurotensin
	Pro-opiocortin		
		Ovine	Somatostatin 1–14
Salmon	Calcitonin		ACTH 1–24
Eel	Calcitonin		
Amphibian	Bombesin		

Table 7   The specificity of HCT radioimmunoassay[52]

Sensitivity	Direct (7d) assay	= 2.0 pg/tube
	Extraction assay	= 0.2 pg/tube
Reproducibility	Direct (7d) assay intra-assay c.v. =	6.4%
	inter-assay c.v. =	8.6%
	Extraction assay intra-assay c.v. =	9.9%
	inter-assay c.v. =	13.4%

Table 8   Some radioimmunoassay characteristics reported for human calcitonin[52]

c.v. = coefficient of variation
d = day

likely responders and non-responders. It has also been used in patients receiving long-term calcitonin treatment to verify that the bone cells (osteoclasts) have not lost their sensitivity to calcitonin (this is known as the 'escape' phenomenon). Other variables, such as plasma phosphate and cyclic AMP, and urinary calcium and cyclic AMP may be measured by this test, in addition to blood calcium.

Other assay methods

Other methods for calcitonin assay which are under development or already in use (mainly for research purposes) include high-pressure liquid chromatography, radioreceptor assay, immunoenzymatic assay, determination via cyclic AMP levels in plasma, the kidneys or urine, and quantitative cytochemistry[58-61].

Units

The potency, or activity, of a calcitonin (natural or synthetic, or contained in a pharmaceutical preparation) is established by bioassay and is expressed in terms of the international unit or IU (previously known as the MRC unit because it was established as the biological standard by the British Medical Research Council). This is now the only unit which is officially recognized, although historically a number of different units have been used (Table 9). It is defined as the quantity of calcitonin which, under precisely defined experimental conditions, will produce a fall in blood calcium level after 1 hour equal to that produced by the contents of one ampoule (or part of an ampoule) of the International Reference Preparation of calcitonin when given by injection to young rats. International Reference Preparations are now available for natural porcine calcitonin (PCT, 1 IU/ampoule), synthetic salmon calcitonin (SCT, 80 IU/ampoule), synthetic human calcitonin (HCT, 1 IU/ampoule) and a synthetic aminosuberic analogue of natural eel calcitonin (Asu[1,7]-eel, ECT, 15 IU/ampoule), and a research

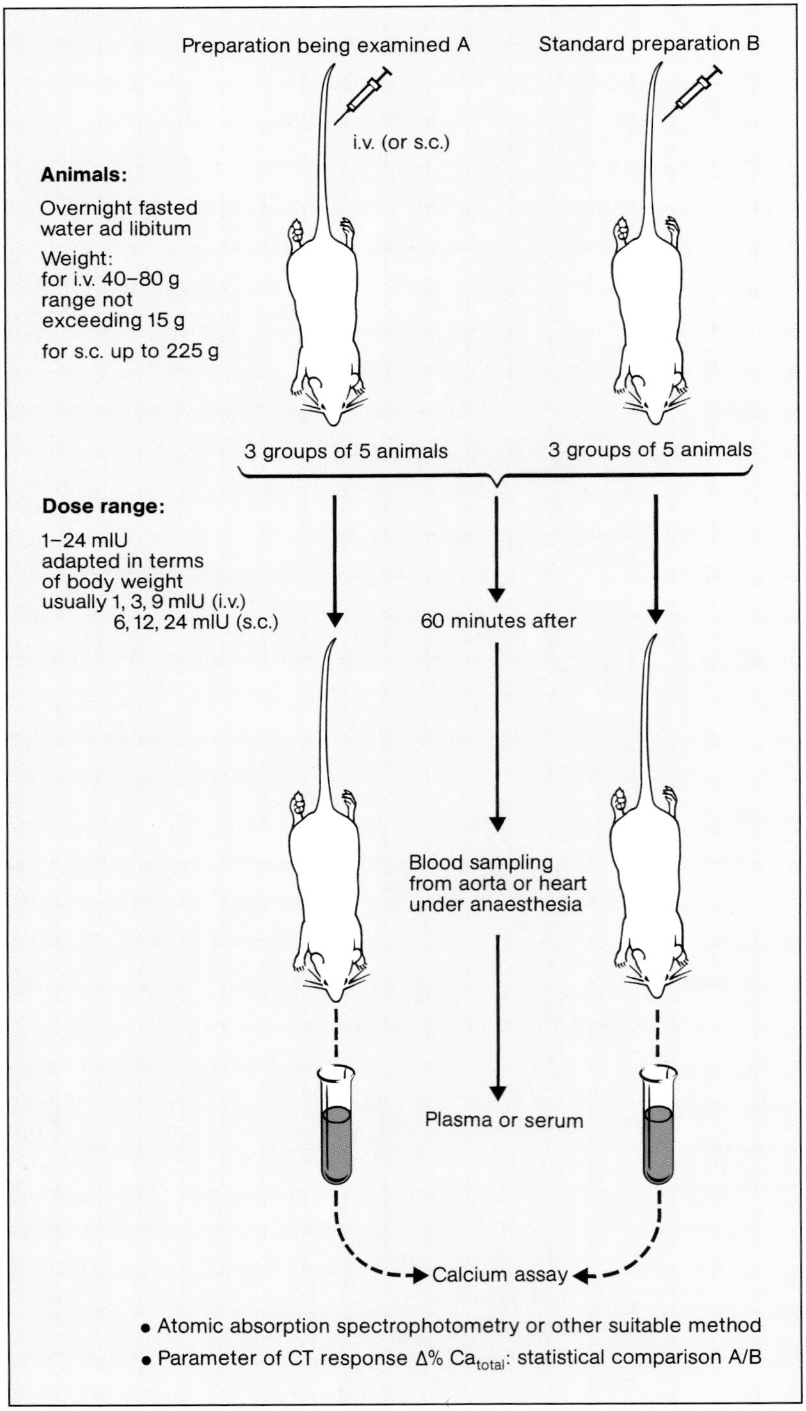

Preparation being examined A

Standard preparation B

i.v. (or s.c.)

Animals:

Overnight fasted
water ad libitum

Weight:
for i.v. 40–80 g
range not
exceeding 15 g

for s.c. up to 225 g

3 groups of 5 animals   3 groups of 5 animals

Dose range:

1–24 mIU
adapted in terms
of body weight
usually 1, 3, 9 mIU (i.v.)
6, 12, 24 mIU (s.c.)

60 minutes after

Blood sampling
from aorta or heart
under anaesthesia

Plasma or serum

Calcium assay

• Atomic absorption spectrophotometry or other suitable method
• Parameter of CT response $\Delta\%\ Ca_{total}$: statistical comparison A/B

Fig. 11   Simplified schema illustrating the
bioassay method for calcitonin

Hirsch unit:	Amount of calcitonin in 0.033 ml of the supernatant of an ultracentrifuged acid extract of porcine thyroid tissue containing 10 μg of nitrogen, as determined by the method of Lowry et al. (62).
Baghdiantz (or Hammersmith) unit:	0.1 of the amount of calcitonin necessary to produce a fall of 0.5 mEq/l (1.0 mg/100 ml, or 10%) in plasma calcium during a 1-hour intravenous infusion in a Hammersmith piebald assay rat weighing 150 g.
Milhaud unit:	Amount of calcitonin producing a fall in serum calcium of 0.1 mg/100 ml within 50−70 minutes after subcutaneous administration to Wistar assay rats weighing 120 g.
MRC unit:	Defined by a stable ampouled preparation of thyrocalcitonin. A crude concentrate of porcine thyrocalcitonin, processed to the salt precipitation stage according to the method of Baghdiantz, was ampouled by the Medical Research Council of London as Research Standard A. Approximately 1000 ampoules were prepared, each containing about 10 mg of extract. The ampoules were assigned a potency of 0.25 units (1 unit = 4 ampoules). There was intended to be sufficient standard material in one ampoule for one *in-vivo* rat hypocalcaemia bioassay, i.e. the MRC unit was approximately equivalent to 1000 Hammersmith units.
	1 MRC U $\approx$ 100 Hirsch U $\approx$ 1000 Baghdiantz U $\approx$ 1000 Milhaud U

Table 9   Obsolete calcitonin units

Definition

1 unit =   Whole or part of the content of the sealed ampoule (reference preparation) calibrated by international experts in a collaborative study and based on hypocalcaemic effect (a standardized bioassay in young rats involving measurement of the fall in serum calcium 1 h after parenteral administration).

The International Unit derives from the early "animal units" and is approximately equivalent to 1000 Hammersmith units or 100 Hirsch units. However there might be variations between animal units, depending on variations between rats and experimental conditions in different laboratories around the world.

International Reference Preparations (IRP)

Prepared by the British National Institute for Biological Standards and Control (NIBSC). Earlier standards were prepared by the Medical Research Council of London.

IRPs are available for:

Porcine calcitonin (code no. 70/306)
Human calcitonin (code no. 70/234)
Salmon calcitonin (code no. 72/158)
[Asu[1,7]]-eel calcitonin (code no. 84/614)

A "research standard" is available for eel calcitonin (code no. 84/547) and an IRP is in preparation.

1 International Unit (IU) = 1 MRC Unit

Equivalence

1 IU PCT $\approx$ 1 IU SCT $\approx$ 1 IU HCT in young rats (hypocalcaemia bioassay) and within the dose range over which the response is linear

UNIT = AMOUNT OF ACTIVITY contained in a defined quantity of calcitonin

One international unit of calcitonin is the quantity which possesses a defined level of activity under standardized experimental conditions. Under these conditions the calcitonins of the various species are approximately equipotent.

N.B. In the future units will probably be defined differently (by weight) since most calcitonins are highly purified and can be described in terms of their physicochemical characteristics, of immunoassay, etc., even though the final identification and calibration will be based on biological effect.

Table 10   Current units of calcitonin activity

standard for synthetic eel calcitonin (Table 10). In the case of synthetic human calcitonin, weight of substance is often used to indicate the content of commercial preparations, although its International Reference Preparation is also expressed in International Units.

In terms of the hypocalcaemic response produced in young rats under identical experimental conditions, the units of potency of the various calcitonins are similar over the dose range producing a linear response. This may also be the case for other species and for different methods of assay, such as the effect on renal cyclic AMP in young rats.

Relative potencies of the calcitonins

The weight of substance which corresponds to 1 unit of calcitonin differs considerably between the various species. As Table 11 shows, the salmon, chicken and Asu[1,7]-eel calcitonins are the most potent weight for weight and the porcine and human calcitonins among the least potent, as confirmed in the induced-hypocalcaemia assay in rats (Fig.12).

The potency of calcitonin preparations depends on their purity. This applies principally to the natural calcitonins, which may contain impurities in the form of low-activity or inactive substance attenuating their 'activeness' and sometimes causing untoward effects. For this and other reasons it is preferable to declare the content of a preparation in units of activity rather than weight of substance.

A number of hypotheses based on experimental findings have been advanced to explain the large differences in potency between the calcitonins[19]:
– That salmon calcitonin is more resistant to enzyme degradation, both *in vivo* and *in vitro*, than human and porcine calcitonin[64].
– That the metabolic clearance rate (MCR) of salmon calcitonin is lower than that of human and porcine calcitonin[65].

	SCT		HCT		PCT
MCR ratio	1	:	2.2	:	3.1
			1.0	:	1.4

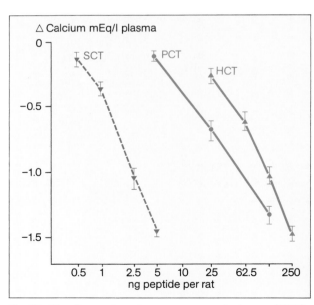

Fig. 12   Dose/hypocalcaemic response curves for human (HCT), pig (PCT) and salmon (SCT) calcitonins, as determined 50 minutes after a single intravenous injection in young rats. Each point represents the mean of 10 rats ± SEM[19,63].

Species	Activity (units/mg)	Activity (mg/unit)
Salmon I* Salmon II Chicken Asu[1,7]-eel	4000–6000	0.00017–0.00025
Salmon III Eel	2000–4000	0.00025–0.0005
Rat	$\simeq$ 400	$\simeq$ 0.0025
Ox Sheep Pig Man	100–200	0.005–0.01

Table 11   Hypocalcaemic activity of the calcitonins (adapted[19])

*The form discussed in this book

As determined by bioassay	SCT	ECT	HCT	PCT	Ref.
In vitro					
Rat plasma/serum (37°C)	+++	+++	++	+	64
Human plasma	+++			+	
Rat liver/kidney (homogenate)	+++	+++	++	+	
Affinity for membrane receptors of bone/kidney	+++		++	++	
In vivo					
Duration of action	+++	+++	++	+	19
Half-life (normal/nephrectomized rat)	+++	+++	++	+	

Table 12   Biological stability of the four principal calcitonins

+++ = long duration
➔ + = short duration of activity

– That salmon calcitonin and the aminosuberic analogue of eel calcitonin differ from the other calcitonins in having hydrophilic properties which facilitate hydrogen bonding, and in having both more leucine residues and having them in a helicoid configuration, and that these differences equip them better than human and porcine calcitonin to fit the bone and kidney cell receptor sites, i.e. give them greater affinity[19] (Table 12).

Structure-activity relationship[19,20,66–70]

The biological activity of the calcitonins seems to depend on the integrity of the molecule both in shape and 'size' (all 32 aminoacid residues must be present). However:

– No specific site of activity has been identified.

– Full activity appears to depend in some way on the presence of a 7-residue ring at the N terminal, although the disulphide bridge (–S–S–) is not indispensable and may be opened or replaced by some other form of link (–C–C– in the case of the aminosuberic analogue of eel calcitonin[18]).

On the other hand, it is beginning to appear as though quite significant modification of the 'basic' structure might be possible without major loss of hypocalcaemic effect. However, such work is still in the early stages and at the present time only limited modification of the molecule is possible without loss of activity – for example, structural changes at position 8 and at the C terminal, and the introduction of iodine (iodination being very useful for radioimmunoassay purposes), at least in the case of salmon calcitonin[71]. It also appears that, whereas the potency of a low-activity calcitonin such as human can be improved by modifying its structure to make it more similar to a particularly active form such as salmon calcitonin (e.g work by Maier, Riniker, Rittel et al. as reviewed in [19]), no synthetic analogue has yet been produced with a potency much greater than that of natural fish calcitonins. Recent claims that a "superagonist analogue" of salmon calcitonin (homoserine[31]-SCT I) with 10–15 times the hypocalcaemic potency of native hormone has been synthesized by a recombinant DNA method[72] are probably exaggerated, although 2–3 times might be a possibility.

References

1   Baber EC: Contributions to the minute anatomy of the thyroid gland of dog. Proc R Soc Lond 1876, 24, 240–1.

2   Zondek H, Ucko H: Über die wirksame Substanz der Epithelkörperchen. Klin Wschr 1966, 44, 528.

3   Nonidez JF: The origin of the 'parafollicular' cell, a second epithelial component of the thyroid gland of the dog. Am J Anat 1932, 49, 479–505.

4   Copp DH, Davidson AGP: Direct Humoral Control of Parathyroid Function in the Dog. Proc Soc exp Biol Med 1961, 107, 342–4.

5   Copp DH et al: Evidence for a new parathyroid hormone which lowers blood calcium. Proc Canad Fed Biol Soc 1961, 4, 17.

6   Copp DH et al: Evidence for Calcitonin – A New Hormone from the Parathyroid That Lowers Blood Calcium. Endocrinology 1962, 70, 638–49.

7   Kumar MA, Foster GV, MacIntyre I: Further evidence for calcitonin. A rapidly-acting hormone which lowers plasma calcium. Lancet 1963, 2, 480–2.

8   Hirsch PF et al: Thyrocalcitonin: Hypocalcemic Hypophosphatemic Principle of the Thyroid Gland. Science 1964, 146, 412–3.

9   Foster GV, Baghdiantz A, Kumar MA, Slack E, Soliman HA, MacIntyre I: Thyroid origin of calcitonin. Nature 1964, 202, 1303–5.

10 Foster GV, MacIntyre I, Pearse AGE: Calcitonin production and the mitochondrion-rich cells of the dog thyroid. Nature 1964, 203, 1029–30.

11 Pearse AGE: The cytochemistry of the thyroid C cells and their relationship to calcitonin. Proc R Soc Lond (Biol) 1966, 164 (996), 478–87.

12 Pearse AGE: The Thyroid Parenchymatous Cells of Baber, and the Nature and Function of their C Cell Successors in Thyroid, Parathyroid and Ultimobranchial Bodies. In: Calcitonin. Proceedings of the Symposium on Thyrocalcitonin and the C Cells. Ed. S Taylor. Heinemann Medical Books, London, 1968, 98–109.

13 O'Dor RK, Parkes CO, Copp DH: Amino acid composition of salmon calcitonin. Can J Biochem 1969, 47, 823–5.

14 Niall HD, Keutmann HT, Copp DH, Potts JT Jr: Amino acid sequence of salmon ultimobranchial calcitonin. Proc Natl Acad Sci USA 1969, 64, 771–8.

15 Guttmann S et al: Synthese von Salm-Calcitonin, einem hochaktiven hypocalcämischen Hormon. Helv chim Acta 1969, 52, 1789–95.

16 Neher R et al: Human calcitonin. Nature 1968, 220, 984–6.

17 MacIntyre I, Craig RK: Molecular evolution of the calcitonins. In: Neuropeptides: Basic and clinical aspects. Proceedings of the eleventh Pfizer International Symposium, September 1981. Ed. Fink G, Whalley LJ, Churchill Livingstone 1982, 255–8.

18 Morikawa T, Munekata E, Sakakibara S, Noda T, Otani M: Synthesis of eel-calcitonin and Asu^{1,7}-eel-calcitonin: Contribution of the disulfide bond to the hormonal activity. Experientia 1976, 32, 1104–6.

19 Guttmann S: Chemistry and structure-activity relationship of natural and synthetic calcitonins. In: Calcitonin 1980. Proc Int Symp, Milan 1980. Ed. Pecile A. Excerpta Medica Int Congr Ser 1981, 540, 11–24.

20 Neher R et al: Struktur von Calcitonin M und D. Helv chim Acta 1968, 51, 1900–5.

21 Riniker B et al: Menschliches Calcitonin – Isolierung und Charakterisierung. Helv chim Acta 1968, 51, 1738–42.

22 Girgis SI et al: An immunological comparison of normal circulating calcitonin with calcitonin from medullary carcinoma. In: Molecular endocrinology. Ed MacIntyre I, Szelke M. Elsevier, 1977, 175–8.

23 Amara SG, David DN, Rosenfeld MG, Roos BA, Evans RM: Characterization of rat calcitonin MRNA. Proc Natl Acad Sci USA 1980, 77, 4444–8.

24 Amara SG et al: Expression in brain of a messenger RNA encoding a novel neuropeptide homologous to calcitonin gene related peptide. Science 1985, 229, 1094–7.

25 Rosenfeld MG, Mermod JJ, Amara SG, Swanson LW, Sawchenko PE, Rivier J, Vale WW, Evans RM: Production of a novel neuropeptide encoded by the calcitonin gene via tissue-specific RNA processing. Nature 1983, 304, 129–35.

26 Allison J, Hall L, MacIntyre I, Craig RK: The construction and partial characterization of plasmids containing complementary DNA sequences to human calcitonin precursor polyprotein. Biochem J 1981, 199, 725–31.

27 Craig RK, Hall L, Edbrooke MR, Allison J, MacIntyre I: Partial nucleotide sequence of human calcitonin precursor MRNA identifies flanking cryptic peptides. Nature 1982, 295, 345–7.

28 MacIntyre I, Hillyard CJ, Murphy PK, Reynolds JJ, Das RE, Craig RK: A second plasma calcium-lowering peptide from the human calcitonin precursor. Nature 1982, 300, 460–2.

29 Roos BA, Fischer JA, Pignat W, Alander CB, Raisz LG: Evaluation of the in vivo and in vitro calcium-regulating actions of noncalcitonin peptides produced via calcitonin gene expression. Endocrinology 1986, 118, 46–51.

30 Brain SD, MacIntyre I, Williams TJ: A second form of human calcitonin gene-related peptide which is a potent vasodilator. Eur J Pharmacol 1986, 124, 349–52.

31 Brain SD, Williams TJ, Tippins JR, Morris HR, MacIntyre I: Calcitonin gene-related peptide is a potent vasodilator. Nature 1985, 313, 54-6.

32 Terenghi G, Polak JM, Ghatei MA, Mulderry PK, Butler JM, Unger WG, Bloom SR: Distribution and origin of calcitonin gene-related peptide (CGRP) immunoreactivity in the sensory innervation of the mammalian eye. J Comp Neurol 1985, 233, 506–16.

33 Dawbarn D, Gregory J, Emson PC: Visualization of calcitonin gene-related peptide receptors in the rat brain. Eur J Pharmacol 1985, 111, 407–8.

34 Gibson SJ et al: Calcitonin gene-related peptide immunoreactivity in the spinal cord of man and of eight other species. J Neurosci 1984, 4, 3101–11.

35 Lundberg JM, Franco Cereceda A, Hua X, Hokfelt T, Fischer JA: Co-existence of substance P and calcitonin gene-related peptide-like immunoreactivities in sensory nerves in relation

to cardiovascular and bronchoconstrictor effects of capsaicin. Eur J Pharmacol 1985, 108, 315–9.

36   Tschopp FA, Tobler PH, Fischer JA: Calcitonin gene-related peptide in the human thyroid, pituitary and brain. Mol cell Endocrinol 1984, 36, 53–7.

37   MacIntyre I: The calcitonin gene peptide family and the central nervous system. In: Endocrinology. Proc 7th int Congr Endocrinol, Quebec City July 1984. Eds Labrie F, Proulx L. Excerpta Medica Int Congr Ser 1984, 655, 930–3.

38   Tschopp FA, Henke H, Petermann JB, Tobler PH, Janzer R, Hokfelt T, Lundberg JM, Cuello C, Fischer JA: Calcitonin gene-related peptide and its binding sites in the human central nervous system and pituitary. Proc Natl Acad Sci USA 1985, 82, 248–52.

39   Inagaki S, Kito S, Kubota Y, Girgis S, Hillyard CJ, MacIntyre I: Autoradiographic localization of calcitonin gene-related peptide binding sites in human and rat brains. Brain Res 1986, 374, 287–98.

40   Girgis SI, MacDonald DW, Stevenson JC, Bevis PJ, Lynch C, Wimalawansa SJ, Self CH, Morris HR, MacIntyre I: Calcitonin gene-related peptide: Potent vasodilator and major product of calcitonin gene. Lancet 1985, 2, 14–6.

41   Morris HR, Panico M, Etienne T, Tippins J, Girgis SI, MacIntyre I: Isolation and characterization of human calcitonin gene-related peptide. Nature 1984, 308, 746–8.

42   Tippins JR, Morris HR, Panico M, Etienne T, Bevis P, Girgis S, MacIntyre I, Azria M, Attinger M: The myotropic and plasma-calcium modulating effects of calcitonin gene-related peptide (CGRP). Neuropeptides 1984, 4, 425–34.

43   Fisher LA, Kikkawa DO, Rivier JE, Amara SG, Evans RM, Rosenfeld MG, Vale WW, Brown MR: Stimulation of noradrenergic sympathetic outflow by calcitonin gene-related peptide. Nature 1983, 305, 534–6.

44   Gennari C, Fischer JA: Cardiovascular action of calcitonin gene-related peptide in humans. Calcif Tiss Int 1985, 37, 581–4.

45   Struthers AD et al: The acute effect of human calcitonin gene-related peptide in man. J Endocr 1985, 104, Suppl 129.

46   Edbrooke MR et al: Expression of the human calcitonin/CGRP gene in lung and thyroid carcinoma. EMBO J 1985, 4, 715–24.

47   Cohen SL, MacIntyre I, Grahame-Smith D, Walker JG: Alcohol-stimulated calcitonin release in medullary carcinoma of the thyroid. Lancet 1973, 2, 1172–3.

48   Lenz HJ et al: Calcitonin gene-related peptide acts within the central nervous system to inhibit gastric acid secretion. Regulatory Peptides 1984, 9, 271–7.

49   Lenz HJ, Mortrud MT, Rivier JE, Brown MR: Calcitonin gene-related peptide inhibits basal, pentagastrin, histamine, and bethanecol stimulated gastric acid secretion. Gut 1985, 26, 550–5.

50   Zanelli JM, Lane E, Kimura T, Sakakibara S: Biological activities of synthetic human parathyroid hormone (PTH) 1–84 relative to natural bovine 1–84 PTH in two different in vivo bioassay systems. Endocrinology 1985, 117, 1962–7.

51   Copp DH, Ma SWY: Comparative endocrinology of calcito-

nin. In: Calcitonin 1980, Proc Int Symp, Milan 1980. Ed. Pecile A. Excerpta Medica Int Congr Ser 1981, 540, 53–66.

52   MacIntyre I et al: Essential steps in the measurement of normal circulating levels of calcitonin. Biomed Pharmacother 1984, 38, 230–4.

53   David L et al: Mise en évidence d'artéfacts introduits dans les dosages radioimmunologiques lors du traitement de sérums par le charbon dans le but de les «déshormoniser». Applications aux dosages radioimmunologiques de la calcitonine et de la parathormone. Pathol Biol (Paris) 1975, 23, 833–4.

54   Deftos LJ, O'Riordan JLH: Problems in radioassays of the calcitropic hormones. In: Endocrinology of Calcium Metabolism, Proc 6th Parathyroid Conf. Ed. Copp DH, Talmage RV. Excerpta Medica Int Congr Ser 1978, 421, 345–8.

55   Straus E, Yalow RS: Artifacts in the radioimmunoassay of peptide hormones in gastric and duodenal secretions. J Lab Clin Med 1976, 87, 292–8.

56   Deftos LJ: Pathophysiology of calcitonin secretion in different species. In: The Effects of Calcitonins in Man, Proc 1st Int Workshop, Florence 1982. Ed. Gennari C, Segre G. Masson, Milan 1983, 43–53.

57   Fournie A et al: Test d'hypocalcémie aiguë à la calcitonine de porc et de saumon. Rev Rhum 1977, 44, 91–8.

58   Marx SJ, Woodward CJ, Aurbach GD: Calcitonin receptors of kidney and bone. Science 1972, 178, 999–1001.

59   Heersche JN, Marcus R, Aurbach GD: Calcitonin and the formation of 3;5'-AMP in bone and kidney. Endocrinology 1974, 94, 241–7.

60   Mahaffey JE et al: Calcitonin receptors on cultured pig kidney cells. Calcif Tiss Int 1979, 28, Suppl 154.

61   Salmon DM et al: Quantitative cytochemical responses to exogenously administered calcitonins in rat kidney and bone cells. Mol cell Endocrinol 1983, 33, 293–304.

62   Lowry OH et al: Protein measurement with the folin-phenol reagent. J biol Chem 1951, 193, 265–75.

63   Maier R: Pharmacology of human calcitonin. In: Human calcitonin and Paget's disease. Proc int Workshop, London 1976. Ed. MacIntyre I. Huber 1977, 66–77.

64   Newsome FE, O'Dor RK, Parkes CO, Copp DH: A study of the stability of calcitonin biological activity. Endocrinology 1973, 92, 1102–6.

65   Nüesch E, Schmidt R: Comparative pharmacokinetics of calcitonins. In: Calcitonin 1980. Proc int Symp, Milan 1980. Ed Pecile A. Excerpta Medica Int Congr Ser 1981, 540, 352–64.

66   Potts JR, Aurbach GD: Chemistry of the calcitonins. In: Handbook of Physiology. Ed Greep RO, Astwood EB. American Physiological Society 1976, 7/7, 423ff.

67   Kumar MA et al: A biological assay for calcitonin. J Endocr 1965, 33, 469–75.

68   Merle M et al: Acylation of porcine and bovine calcitonin: Effects on hypocalcemic activity in the rat. Biochem Biophys Res Comm 1977, 79, 1071–6.

69   Nozaki S: Synthesis of (2-serine, 8-valine)-human calcitonin. Bull chem soc Japan 1978, 51, 2995–3003.

70   Sieber P et al: Synthesis and biological activity of peptide se-

quences related to porcine thyrocalcitonin. In: Calcitonin 1969, Proc 2nd int Symp, London 1969. Ed Taylor SF, Forster G. Heinemann 1970, 28–33.

71   Maurer R et al: Comparison of three ligands for calcitonin binding sites. In: Calcitonin 1984. Selected short communications presented at the international symposium 'Calcitonin 1984' in Milan. Ed Doepfner WEH. Excerpta Medica 1986, 1–9.

72   Heath H III et al: Homoserine[31]-salmon calcitonin I ([HSE[31]]-CT): synthesis of a superagonist of CT by a recombinant DNA method. J Bone Min Res 1986, 1, Suppl 1, Abstr 344.

Chapter 2: Endogenous Calcitonin

Foreword
by Professor Iain MacIntyre

While calcitonin was first discovered in the early 1960s[1-5], it is only within the last few years that actual proof of its direct inhibitory action on osteoclasts has been forthcoming[6-8], although such an action had been predicted soon after its discovery[9–11].

In this section a more comprehensive description is given of the other actions of this interesting peptide. It is clear that calcitonin acts on the kidney, the gut and within the central nervous system and that, although of lesser physiological importance in man, these actions may reflect important roles in more primitive vertebrates. Nevertheless, in order to understand the major action of calcitonin in skeletal conservation, we need to see this against a wider physiological background. I believe this section provides this and will be of use to basic scientists, clinical investigators and physicians.

I. MacIntyre

Sources of secretion

The primary source of endogenous calcitonin secretion is a cell system comprising parafollicular (C) cells, which derive from the embryonal neural crest[3,4,12,13].

These cells, first described by Baber in 1876[14], have a clear nucleus and richly granular cytoplasm and differ histologically and cytochemically from thyroxine-secreting cells. They occur primarily in the thyroid gland in mammals[15] and in the ultimobranchial body in fish, amphibians, reptiles and birds[15-18], but also migrate to other sites during embryonal development to become secondary sources of immunoreactive calcitonin, or iCT (Fig. 13). In the lizard for example, the lung as well as the ultimobranchial body is an important source of calcitonin[18], and in man (but not in the rat) C cells are present in the lungs and thymus as well as in the thyroid[19]. This explains why calcitonin has been found in both the blood and urine of patients after thyroidectomy and in the organs of healthy monkeys 2 months after removal of the thyroid[20] (Table 13). It may also explain why medullary thyroid carcinomas sometimes metastasize to the thymus.

Immunoreactive calcitonin has also been found in many other tissues, in both man[19] (Table 14) and in animals[19-24]. One such site is the central nervous system, where it probably acts as a neuromodulator[23,24]. A substantial amount of iCT is sometimes found bound to receptors in organs such as the liver, which may thus function as reservoirs from which calcitonin is liberated in response to various stimuli, such as those connected with food intake[20].

Tissue	Ab-IV* (pg/g wet weight)	
	Intact	After thyroidectomy
Thyroid	396000 ± 83000	–
Liver	12390 ± 350	478 ± 161
Thymus	1640 ± 370	1780 ± 910
Lung	596 ± 438	982 ± 1116
Submaxillary gland	842 ± 27	836 ± 127
Parotid gland	759 ± 554	490 ± 13
Duodenum	751 ± 110	490 ± 115
Jejunum	463 ± 80	415 ± 134
Stomach	420 ± 139	335 ± 125
Skeletal muscle	266 ± 35	308 ± 115
Kidney	241 ± 2	285 ± 58
Hypothalamus	48 ± 11	457 ± 791

Table 13   Tissue levels of immunoreactive calcitonin in intact monkeys and in monkeys 2 months after thyroidectomy (mean ± SD) (adapted[20])

*Ab-IV = antiserum able to recognize fraction 26−32 (= carboxyl terminal)

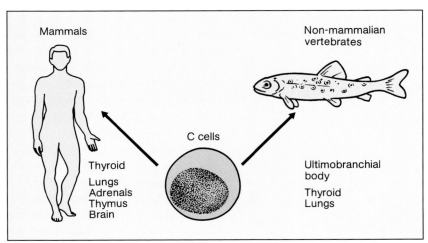

Mammals

Non-mammalian vertebrates

C cells

Thyroid
Lungs
Adrenals
Thymus
Brain

Ultimobranchial body
Thyroid
Lungs

Fig. 13   Principal and secondary sites of calcitonin-secreting cells in mammals and non-mammalian vertebrates

Site	Site
Thyroid	Cerebral cortex
Jejunum	Gall bladder
Thymus	Kidney
Bladder	Adipose tissue
Lung	Adrenals
Rectum	Pancreas
Testicle	Spleen
Muscle (skeletal)	Heart
Hypothalamus	Cerebellum
Pituitary	Ovary
Lymph node	Cardiac blood
Liver	Erythrocytes
Stomach	Bone
Oesophagus	Seminal plasma

Table 14 Sites at which calcitonin has been detected in human tissues by means of an antibody (Ab-IV) capable of recognizing its carboxyl terminal (adapted[20])

Bronchial carcinoid	Renal clear-cell cancer
Intestinal carcinoid	Testicular dysembryoma
Phaeochromocytoma	Myeloma
Stroma ovarii	Oesophageal cancer
Mucosal neuroma	Pleural epithelial cancer
Bronchial carcinoma	Thymic carcinoid
Melanoma	Colonic cancer
Insulinoma	Pancreatic cancer
Hepatoma	Gastric cancer
Mammary cancer	Vipoma
Ganglioneuroma	Pancreatic somatostatinoma
Sympathoblastoma	Paraganglioma
Laryngeal apudoma	

Table 15 Ectopic sources (tumours) of calcitonin (listed in chronological order of discovery) as detected by either bioassay, RIA or immunofluorescence (adapted[25])

Three hypotheses have been advanced to explain the persistence of iCT after thyroidectomy:

- That it is receptor-bound calcitonin originating from the thyroid before excision.

- That calcitonin is still being secreted by residual functioning thyroid tissue or by an extrathyroidal system of parafollicular cells such as the bronchial K cells.

- That the calcitonin gene has been incompletely suppressed[20].

Although medullary thyroid carcinomas are the source of the most profuse iCT secretion, causing blood levels as much as 5000 times higher than normal, iCT is also secreted by many other types of neoplasm[25] (Table 15). This makes it a useful marker for diagnosis and for monitoring the effectiveness of treatment in such cases[26-30]. Its production is probably due to the "derepression" of APUD cells – to which the C cells belong – in such tumours.

Calcitonin receptors

The identification of specific binding sites for a hormone is often a valuable pointer to the identity of its target cells and organs and to the nature of its biological activity. The transmission of messages within the cell – the mechanism by which biological effects are mediated – frequently involves interaction with a membrane receptor coupled to adenylate cyclase and a change in the intracellular concentration of cyclic AMP. Such changes can consequently be used to assess a hormone's activity.

Research to identify calcitonin receptors started in about 1970 with binding and autoradiographic studies. Binding sites, or receptors, have been discovered in many different tissues, including bone and kidney cells, the central nervous system and the pituitary gland, lymph cells and some types of tumour cells[23,31–46] (Table 16). Their presence in bone and kidney is evidence for the role of calcitonin in bone and renal metabolism, but its precise function in the CNS and pituitary has yet to be elucidated. At all events,

Bone (38)	Osteoclasts and some marrow cells
Kidney (39, 40, 54, 55)	Proximal and distal convoluted tubules Ascending limb of the loop of Henle Cortical segment of the collecting tubule
CNS and pituitary (23, 34, 36, 37, 41)	**High density of receptors** Hypothalamus Preoptic nucleus and nucleus accumbens Amygdaloid nucleus Zona incerta Interpeduncular nucleus Nucleus griseus periventricularis Reticular formation **Average or low density** Arcuate nucleus and nucleus supra-mamillaris Substantia nigra Pituitary (pars intermedia and anterior lobe) Spinal cord?
Other sites (32, 33, 42–46)	Leydig cells Lymph cells (man) Mammary and bronchial tumour cells

Table 16   Principal tissues and organs in rat and man containing calcitonin-binding sites, or receptors

binding sites for calcitonin are fairly widely distributed, suggesting that it may well have a more important and complex function than what current evidence suggests is a role in regulating calcium homoeostasis. This work has also shown that the calcitonin having the greatest affinity for all known calcitonin receptors is the synthetic salmon peptide and, probably, eel calcitonin and its aminosuberic analogue. This is consistent with the high hypocalcaemic potency of these forms of the hormone.

Bone receptor sites

Calcitonin receptor sites have been identified on osteoclasts[6–8] and they are probably also present on osteocytes and certain marrow cells (monocytes, macrophages, histiocytes, lymphocytes)[38,47–49]. Bind-

ing of calcitonin to the receptor initiates a series of events, which may be represented diagrammatically as follows:

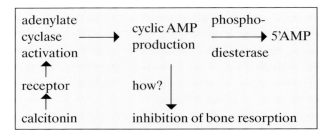

The 'escape' or 'plateau' phenomenon encountered both *in vitro* and in the therapeutic use of calcitonin[50–53] has been attributed to a change in the number of receptors or in their sensitivity, to occupation of the receptors, and even to the emergence of a new population of calcitonin-resistant osteoclasts.

Kidney receptor sites

The presence of calcitonin receptors has been demonstrated in membranes isolated from the renal cortex[35,54] and in human and rodent nephrons (Fig. 14). These receptors are coupled to adenylate cyclase and are distinct from the parathyroid hormone receptors[31,35,39,40]. In the nephron, receptors are located in the ascending limb of the loop of Henle, in the proximal end of the distal convoluted tubule (and to a lesser extent in its distal extremity) and in the cortical segment of the collecting tubule[39,40]. Data on the location and concentration of ^{125}I-SCT binding sites in rat kidney sections from another study[55] are given in Table 17.

CNS and pituitary receptor sites

Specific calcitonin binding sites have been found at various locations in the human[23] (Fig. 15) and

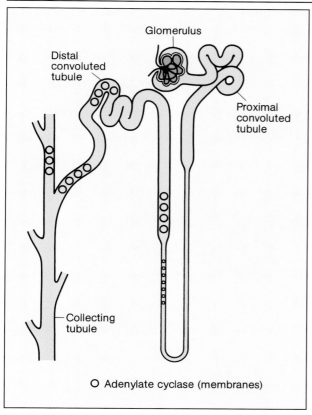

Fig. 14   Calcitonin receptor sites in the human nephron

	Structures	^{125}I-SCT-binding sites (fmol/mg protein)
Cortex	PCT, DCT, CCT (superficial cortex)	17.26 ± 3.26
	Cortical (TAL)	8.52 ± 0.98
Outer medulla	Outer zone (TAL, pars recta)	34.07 ± 1.88
	Middle zone (pars recta, TDL, TAL)	17.25 ± 1.43
	Inner zone (TDL, TAL)	44.60 ± 2.00
Inner medulla	Thin ascending limb	2.49 ± 0.21
Papilla	–	N.D.

Table 17   Location and concentration of ^{125}I-SCT-binding sites in rat kidney sections[55]

Concentrations, expressed as specific binding, were measured by quantitative autoradiography. PCT = proximal convoluted tubule; DCT = distal convoluted tubule; CCT = cortical collecting tubule; TAL = thick ascending limb; TDL = thin descending limb; N.D. = non-detectable

Lymph and human tumour-cell receptor sites

Calcitonin receptors are also present on human lymph cells[42,43] and on mammary[44,45,57] and bronchial tumour cells[46], their characteristics being similar to those of the bone and kidney cell sites, viz.:

– Binding activates adenylate cyclase and the production of cyclic AMP.

– The affinity of calcitonin for these receptor sites is of the same order as for those in bone and kidney cells.

– Loss of sensitivity ('escape phenomenon') occurs in the same way.

Although some tumours, e.g. bronchial, do show ectopic production of several peptides, including calcitonin, the presence of receptors in tumour tissue does not necessarily mean that they are present in normal tissue of the type giving rise to the tumour[46].

rat[24,34,36,37,41] central nervous system and pituitary gland (anterior and intermediate lobes in the case of rat) in studies carried out with ^{125}I-SCT, which is irreversibly bound to the receptors[56]. Their distribution suggests that endogenous calcitonin may have a neuroendocrine, analgesic or more general neuromodulator role at cerebral level[37]. Exogenous calcitonin might have a similar effect, though whether it reaches the brain by crossing the blood-brain barrier has yet to be shown.

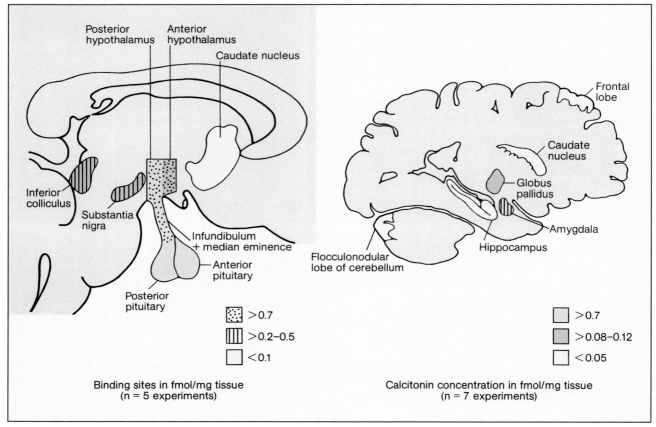

Fig. 15   Concentration of calcitonin and distribution of specific ^{125}I-calcitonin binding sites in human brain and pituitary tissue (homogenates derived from 0.63, 2.5 or 10 mg of fresh tissue)[23]

However, receptors might be present in normal mammary tissue, their function being to regulate calcium transport in the breast during lactation[44].

Blood levels

Normal basal levels

Calcitonin, like many other endogenous hormones, is derived from a precursor. It is present in the blood in several forms, most of which (those with a high molecular weight) are probably inactive[58–60]. The only biologically active form appears to be a monomer which is identical to synthetic human calcitonin. It is broken down in the peripheral blood into inactive fragments, although the detection has been reported[61] in human thyroid and brain of calcitonin-like immunoreactivity of a type resembling salmon calcitonin.

Blood levels can be determined by radioimmunoassay, the specificity of measurement depending on the characteristics of the antibody used. The need for specific antibodies and the difficulty of the techniques involved account for the large discrepancies between reported figures[62], particularly for normal basal levels of endogenous calcitonin in healthy subjects (Fig. 16; Table 18). Expressed as weight per unit volume of *immunoreactive* calcitonin rather than as units of biological activity, quoted levels range from 1 to 500 pg/ml plasma[52,62,71–73]. However, levels are below 100 pg/ml in 75% of subjects[52], the mean level being between 30 and 90 pg/ml[63,65]. A healthy human thyroid contains between 1 and 100 μg calcitonin and the quantity secreted daily is reported as 50–250 μg[67], or 13.8 μg (=1.4 IU) by other authors[68]. It has also been reported that basal levels of extractable endogenous HCT are very low (< 10 pg/ml) in most healthy adults, with speculation that this might correspond to the concentration of circulating monomeric calcitonin[66] (Table 18). However, the biological activity of what is left after extraction is not known.

In view of the apparently wide normal range, blood levels should always be determined using the same assay technique (with or without extraction) and with antiserum of the same specificity. The course of the blood level curve over time is more instructive than a single assay.

Factors that may affect the blood level

Although a given individual's basal calcitonin level normally remains fairly constant, it can be affected by physiological, pharmacological and pathological factors.

● Physiological factors (Table 19)

Calcitonin secretion is affected by a number of physiological factors, most notably by 'calcium status', age and sex-hormone levels. Often, of course, the blood concentration reflects the interplay of several different factors, including PTH (Fig. 17).

The blood level of ionized calcium
This variable, which reflects the rate of bone turnover, is the principal physiological factor regulating the secretion of calcitonin. When the ionized calcium level rises, the secretion and release of calcitonin are stimulated; when it falls they are inhibited[74–76] (Fig. 17).

Food intake
The body responds to food intake by increasing the quantity of calcitonin available in anticipation of a higher calcium load several hours later. It does this by a mechanism closely related to that regulating the secretion of the hormones of the gastrointestinal tract (gastrin, cholecystokinin-pancreozymin, glucagon, caerulin, secretin), which also stimulate calcitonin secretion[52,77,78]. There is some evidence that calcitonin secretion follows a circadian rhythm with a peak between 12 h and 13 h[65], but other evidence suggests that there is no diurnal variation in serum concentrations of either calcitonin or $1,25(OH)_2D_3$[79,80] (Fig. 18).

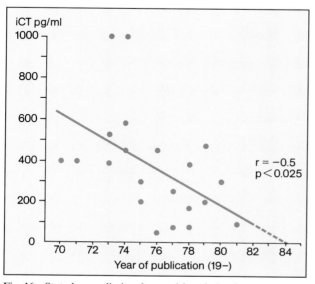

Fig. 16   Stated upper limits of normal for whole-plasma immunoreactive calcitonin (iCT), plotted as a function of publication dates (21 papers). Extrapolation of the regression line suggests that blood levels reached zero in the year 1984![62]

Variable		Reference
Blood		
Mean concentration (in 75% of subjects)	≤100 pg/ml (= 0.02 mlU/ml)*	52
Range of mean values	30−90 pg/ml (= 0.006−0.018 mlU/ml) <10 pg/ml (= <0.002 mlU/ml) extractable iCT	52,63−66
N.B.		
Quantity secreted daily	50−250 μg (=10-50 IU) 13.8 μg (=1.4 IU)** 22 (female) − 59 (male) ng/kg	67 68 69
Quantity present in a normal human thyroid	1−100 μg (= 0.2−20 IU)	70
Cerebrospinal fluid		
Mean concentration	11.1± 1.3 (<2−55) pg/ml (= 2.2 ± 0.26 [<0.4−11] x10^{-3} mIU/ml)	64

Table 18   Normal basal blood and CSF levels of endogenous calcitonin (immunoreactive) in healthy human subjects

* Based on an assumed biological activity of 200 IU/mg
**Figure and conversion as stated in [68]

Factor	Effect on secretion	References
Circadian rhythm	Rhythmic pattern with peri-prandial peak at mid-day	65
Blood level of calcium (and related) ions	Principal regulator of secretion	76−78
Food intake		
− Food rich in calcium and protein	↑ (declining with age)	52,79,80 (81,82)
− Low-calcium diet	↓	81
− Alcohol	↑(single shot)	65
Age		
− Infancy	↑	83
− Ageing process	↓ (no effect)	63 (69)
Sex	Females secrete less than males	63,69,85
	No effect	86
− Pregnancy (esp. in multiparity)	↑	88,89
− Lactation	↑	90,91
− Raised sex hormone levels	↑	86
Raised levels of other hormones		
− Hormones of the GI tract	↑	
− Gastrin	↑	
− Cholecystokinin-pancreozymin	↑	as reviewed in 52
− Glucagon	↑	
− Caerulein	↑	
− Secretin	↑	
− PTH	↑ (indirect)	
− Thyroxine, iodine	↑ (indirect)	63
− Vitamin D ➔ 1,25(OH)$_2$D$_3$	↑	
− Prolactin	↑↓ (via CNS effect)	
Stress	↑	92

Table 19   Physiological factors affecting basal calcitonin secretion in man

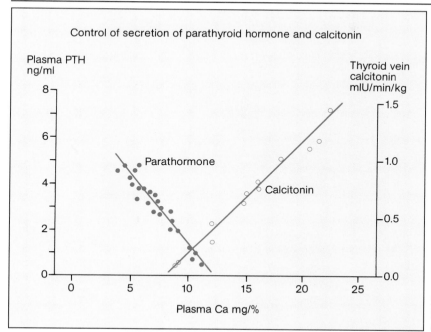

Fig. 17 Effect of plasma calcium on plasma levels of parathyroid hormone and calcitonin secretion rate[74]

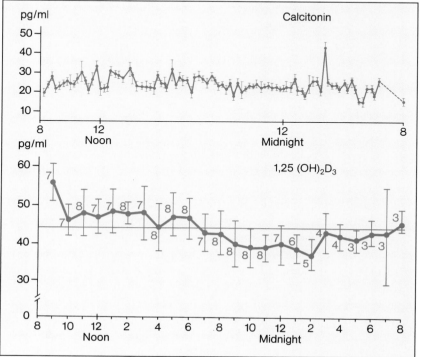

Fig. 18 Mean (± SEM) concentrations of calcitonin and 1,25-dihydroxy-vitamin D$_3$ [1,25(OH)$_2$D$_3$] over 24 hours in 8 adults, as measured by RIA. For calcitonin, assay samples were taken every 15 min, while for 1,25(OH)$_2$D$_3$ determination samples were pooled and assayed at hourly intervals. The figures adjacent to each plot on the 1,25(OH)$_2$D$_3$ curve indicate the number of subjects included[80].

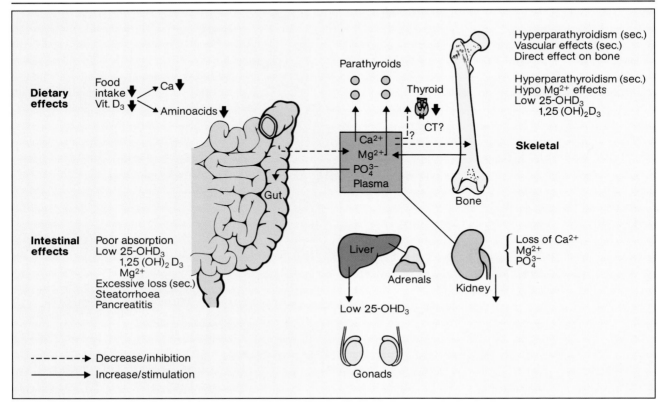

Dietary effects

Food intake ↓
Vit. D₃ ↓
Ca ↓
Aminoacids ↓

Parathyroids

Thyroid
CT?

Hyperparathyroidism (sec.)
Vascular effects (sec.)
Direct effect on bone

Hyperparathyroidism (sec.)
Hypo Mg²⁺ effects
Low 25-OHD₃
1,25 (OH)₂D₃

Ca^{2+}
Mg^{2+}
PO_4^{3-}
Plasma

Gut

Bone

Skeletal

Intestinal effects

Poor absorption
Low 25-OHD₃
1,25 (OH)₂ D₃
Mg²⁺
Excessive loss (sec.)
Steatorrhoea
Pancreatitis

Liver

Adrenals

Kidney

Loss of Ca^{2+}
Mg^{2+}
PO_4^{3-}

Low 25-OHD₃

Gonads

- - - - - ▸ Decrease/inhibition
————▸ Increase/stimulation

Fig. 19   Effects of alcohol on bone metabolism

The quantity of calcitonin secreted depends not only on the calcium content of the food eaten, but also on its protein content. This is also affected by the chronic consumption of alcohol, which exerts an adverse effect of its own on bone metabolism[239] (Fig. 19). A low-calcium diet is associated with low blood levels of calcitonin and the stimulatory effect of food intake on the calcitonin response tends to diminish with age[81,82]. This is probably due either to inadequate intestinal absorption of calcium or to a fall in the quantity of calcitonin available, and is thought to be the explanation for certain bone diseases.

Postprandial calcium regulation and the role of calcitonin are illustrated in Figure 20. Since calcitonin secretion is probably stimulated by food ingestion – particularly if its absorbable calcium content is high – and even by the expectation of eating, increased secretion occurs at particular times over the day in a more or less regular pattern[81]. At the same time, increased calcitonin secretion inhibits the release of calcium from bone into the bloodstream, also giving rise to a slight increase in PTH secretion, which is not inhibited by the ingested calcium. PTH and calcitonin then act in concert to regulate calcium homoeostasis, PTH reducing the amount of calcium excreted in the

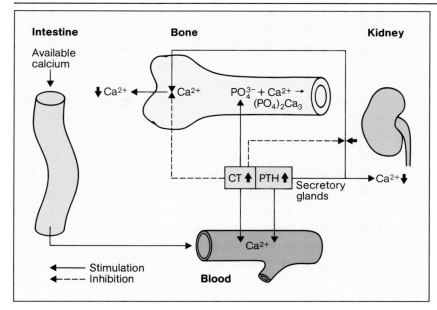

Fig. 20   Regulation of calcium blood levels in response to the dietary intake of calcium

urine and calcitonin suppressing the effect of PTH on calcium efflux from bone.

Calcitonin also actively moves phosphate into the bone fluid and bone lining cells and inhibits its return to the bloodstream. It directs dietary calcium into bone, where it combines with phosphate and is stored as a temporary reservoir. This reserve is the first source to be called upon during, for example, a period of food deprivation, thus reducing the need for PTH-induced bone resorption to maintain calcium balance[81].

Age and sex
In humans the basal level of calcitonin was found to be high at birth and in infancy and to decrease gradually with age[63,83], a sequence which is reversed in rats[84]. Levels appear to be lower in women than in men[85] (Fig. 21) – although this is disputed by some authors[86] – and this difference between the sexes appears to be much greater after the menopause, when basal levels in women fall further. Since calcitonin secretion is stimulated by both oestrogens and testosterone[86], although by an unknown mechanism[87], the postmenopausal fall is attributed to reduced production of these hormones.

On the other hand, it has recently been reported[69] that 148 healthy volunteers showed no appreciable decline with age in basal concentrations of either immunoreactive or extractable calcitonin, or in their secretory capacity in response to calcium infusion. Whole plasma iCT and extractable CT, however, were lower in women than in men both basally (50%) and after calcium stimulation (75%). Secretion rates, based on estimates of plasma monomeric calcitonin from a silica extraction procedure, were also significantly lower for women (22 ± 3 [SEM] ng/day/kg) than men (59 ± 6 ng/day/kg)[69].

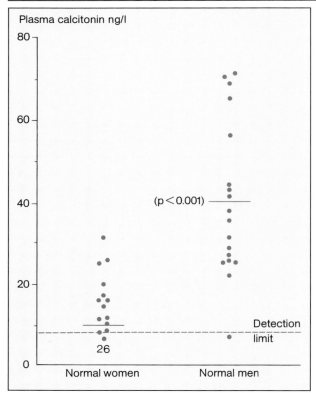

Fig. 21   Plasma calcitonin levels in normal premenopausal females and normal males. In 26 females values were undetectable[85].

Pregnancy

Calcitonin secretion increases during pregnancy, probably as a defence mechanism to protect the maternal skeleton and to meet the additional demands made by the foetus[88,89].

Lactation (Fig. 22)

Lactation is a particularly 'stressful' period as far as the body's calcium requirement is concerned, owing to the simultaneous demands of milk production, of maintaining an adequate and steady blood calcium concentration, and of safeguarding the integrity of the skeletons of both mother and child. To meet these

demands, secretion of three hormones – PTH, calcitonin and $1,25(OH)_2D_3$ – is increased, as demonstrated in rats[90,91]. Calcium turnover rises appreciably in female rats during lactation and basal levels of the three key hormones undergo marked changes. Loss of calcium via the milk leads to a deficiency, which is offset to some extent by enhanced absorption from the diet. Despite this, the calcium balance becomes slightly negative, causing a fall in serum calcium, while levels of PTH, calcitonin and $1,25(OH)_2D_3$ rise. The interplay of these changes protects the maternal skeleton against excessive calcium depletion while ensuring an adequate calcium supply for the offspring. The same hormonal changes also occur in women, though they are thought to be less marked than in rats, in whom the 'stress' of lactation appears to be much more intense than in human beings, presumably because of the larger litter size.

Other physiological factors

Many other factors can affect the blood level of endogenous calcitonin, notably the blood levels of other hormones[52,63,82]. A rise in the blood level of the vitamin-D metabolite $1,25(OH)_2D_3$, for example, increases calcitonin secretion via a direct effect on the C cells. Thus, at times when the concentration of $1,25(OH)_2D_3$ is increased, such as during growth, pregnancy or lactation[89], there is a concomitant rise in the calcitonin level to protect the skeleton and concentrate the action of vitamin D on the gut. It has been postulated that a variety of stressful situations (e.g. physical exertion) are accompanied by an increase in circulating calcitonin (as reviewed in [92]).

The best method of measuring calcitonin levels and detecting deficiencies in either secretion or reserves is a stimulation test using calcium infusion or other agents (see below)[63,93] (Fig. 23). For diagnostic purposes dynamic tests (infusion of a calcium salt inducing qualitative and quantitative variations in the level of endogenous calcitonin) provide more information on the secretory capacity of the C cells and bodily reserves of the hormone.

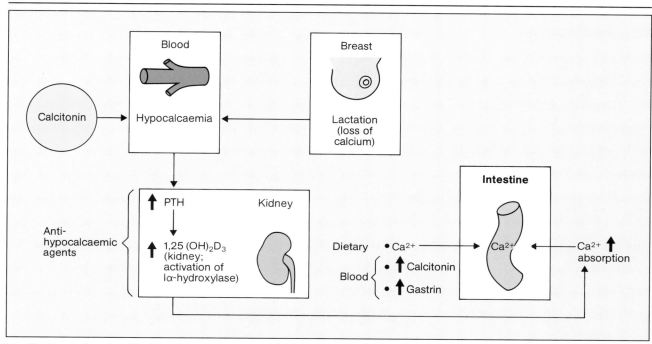

Fig. 22   Regulation of calcitonin, parathyroid hormone, 1,25(OH)$_2$D$_3$ and ionized calcium (Ca^{2+}) blood levels during lactation

● Pharmacological factors

Many pharmacologically active substances can affect blood levels of endogenous calcitonin. Ethanol[65], sulpiride and domperidone[94], pentagastrin[95] and the β-adrenergic agonist isoproterenol[93], for example, cause a rise, whereas cimetidine[94], nicotine[92] and somatostatin[93] have the opposite effect. Dopamine, levodopa, metoclopramide and chlorpromazine[96] have no effect.

Oestrogenic agents are thought, though not universally[97], to increase blood levels of calcitonin[85,87] (Fig. 24) if secretion and storage mechanisms are functioning normally, and incubation studies on rat pup thyroid demonstrated that both 17β-oestradiol and progesterone stimulate calcitonin secretion *in vitro* by

a direct effect on thyroid C cells[98]. This suggests how oestrogen deficiency might be a factor in the pathogenesis of osteoporosis in postmenopausal and ovariectomized women, with low calcitonin levels leading to reduced matrix formation, poor mineralization and removal of the brake on resorption resulting in loss of bone. This is why oestrogens are used, sometimes in combination with calcitonin, to treat postmenopausal and postovariectomy osteoporosis[99–102], in spite of the risks associated with their unwanted effects on the breast, uterus (haemorrhage, risk of endometrial cancer) and cardiovascular system. On the other hand, the dosages given are normally very low and progestagens are often added, so that the risk is reduced[103]. Oral contraceptives are another important category of pharmacological agent with a stimulatory effect on calcitonin secretion[87].

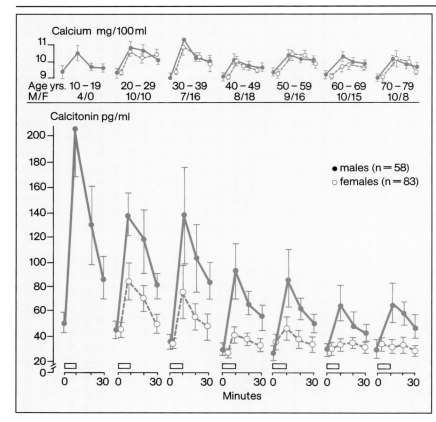

Fig. 23   Response of plasma calcitonin and serum calcium to a 10-minute infusion of calcium in 58 normal men (solid circles) and 83 normal women (open circles). Bars (▭) denote infusion of calcium (as the chloride salt) at a dosage of 3 mg/kg of body weight. Values represent mean ± SEM for each determination. Indicated at the top are the numbers of male and female subjects in each decade. Plasma calcitonin is higher in men than in women ($p < 0.05$ to 0.001) and there is a progressive decrease with age in both sexes[63].

According to one report[104], calcitonin (salmon) itself given by the intravenous route depressed plasma levels of endogenous hormone (Fig. 25), although a recent report[238] states that there was no change in this variable after either intramuscular or intranasal salmon calcitonin.

These and other substances affecting the endogenous production of calcitonin are listed in Table 20.

● Pathological factors

Because of the wide range of apparently normal blood levels of calcitonin, it is sometimes difficult to decide whether a given value is pathological or not. Where this distinction can be made, the calcitonin blood concentration is an excellent biochemical marker both for diagnosing certain diseases and for monitoring the effects of treatment. The best example of this is in medullary carcinoma of the thyroid[52,110–112], where the plasma level falls to below 100 pg/ml during true remission but remains elevated after incomplete excision of the thyroid or when metastases have formed[110,112]. Pathological blood levels may be due to defective secretion of the hormone or to storage malfunction, and the best way of detecting the outcome of either defect is, as has already been said, to use one of the dynamic tests based on provocation by calcium or a hormone such as pentagastrin.

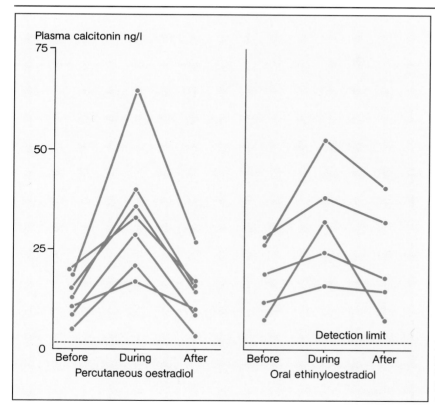

Fig. 24   Plasma calcitonin levels in postmenopausal women before, during and after 12 weeks' treatment with either percutaneous oestradiol ("Oestrogel") or ethinyloestradiol[85].

Factor	References	Factor	References
Substances stimulating secretion		**Substances inhibiting secretion**	
Ethanol	65	Cimetidine	94
Sulpiride	94	Nicotine	92
Domperidone	94	Somatostatin	93
Pentagastrin	95	Calcitonin (exogenous) (?)	104
Isoproterenol	93		
Oestrogens (?)	85, 87, 102 (97)		
Testosterone	87,105	**Substances having no effect**	
Oral contraceptives	87		
Calcium (i.v. infusion)	52,106	Dopamine	96
Catecholamines	92,107	Levodopa	96
Vitamin D	108	Metoclopramide	96
Opioids (in heroin addicts)	109	Chlorpromazine	96
		Bromocriptine	94
		TRH	94

Table 20   Some pharmacological factors affecting basal calcitonin secretion in man

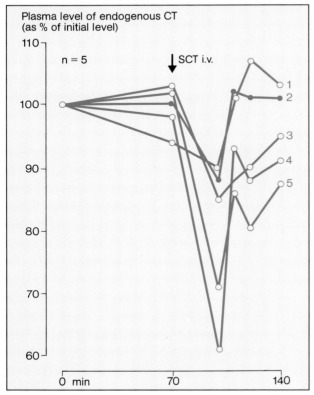

Fig. 25   Decrease in plasma endogenous calcitonin in man after injection of salmon calcitonin[104]

Conditions associated with depressed calcitonin blood levels

Osteoporosis may be associated with calcitonin deficiency, as has been shown to be the case in elderly and non-elderly osteoporotics[113] (Figs 26–27) and in osteoporosis following ovariectomy or the menopause[114]. It may also be true of some types of osteoporosis secondary to endocrine disorders such as hypothyroidism[115] and hypogonadism[105], but there is still too little evidence to support a general connection between these two variables.

Secretion is also depressed in non-goitrous congenital cretinism[116] and, probably, in acromegaly[105]. Thyroidectomy is also followed by low blood levels of calcitonin, of course, provided any ectopic secretion is only slight.

Conditions associated with raised calcitonin blood levels

A number of disorders may give rise to an abnormally high blood level of calcitonin, malignant diseases being the most common:

– Calcitonin-secreting tumours

Such tumours may be thyroid[110] (Fig.28) or extrathyroid (Table 21), though they usually occur in tissue containing C cells, i.e. tissue derived from the neural crest[118], or in tissue belonging to the APUD (Amine Precursor Uptake Decarboxylation) system[119,120] (Table 22). The calcitonin secreted by such tumours may be normal or abnormal in structure. Also, levels as determined by radioimmunoassay may be affected by factors specific to the pathology of the tumour[29,71,121–124]. Release of calcium from bone due to causes such as bone metastases (e.g. from breast tumours) may also lead to raised calcitonin blood levels[125]. The measurement of immunoreactive calcitonin can be useful in diagnosing and in monitoring the management of patients with calcium-secreting tumours such as small-cell carcinoma of the lung[126].

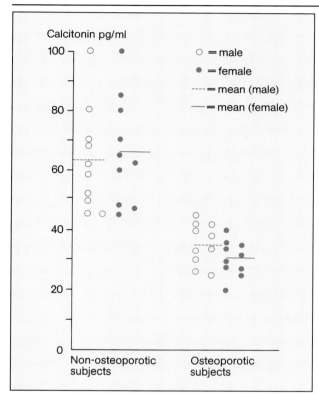

Fig. 26   Basal calcitonin levels in a group of 40 adults with a mean age of 32.7 years[113]

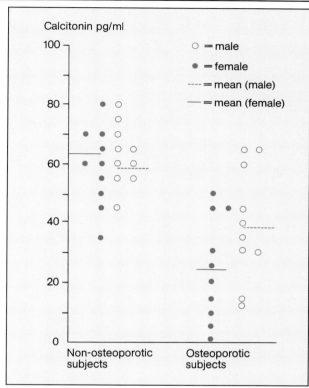

Fig. 27   Basal calcitonin levels in a group of elderly subjects with a mean age of 72.4 years[113]

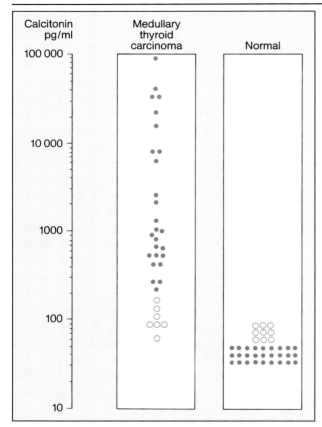

Fig. 28   Basal plasma calcitonin measurements in 33 subjects with histologically confirmed medullary thyroid carcinoma and in 36 normal subjects. At left (closed circles), patients with elevated basal concentrations of calcitonin; at right (closed circles), normal patients with undetectable concentrations of calcitonin. Some patients with thyroid tumours (open circles, at left) had basal calcitonin levels that could not be clearly distinguished from some normal subjects (open circles, at right)[117].

Site	Type of tumour	Reference
Thyroid	Medullary carcinoma of the thyroid with hyperplasia or prehyperplasia of the C cells	110, 111
	Acid-mucopolysaccharide-type trabecular cancer	127
	Trabecular cancer of the thyroid	112
	C-cell adenoma	26
Extrathyroid (ectopic)	Intestinal carcinoid	26, 128
	Bronchial carcinoid, esp. small-cell carcinoma	28, 125 126, 129
	Bone metastases	112
	Melanoma	130
	Breast cancer	130
	Phaeochromocytoma	26, 131
	Hepatoma	25
	Myeloma	25
	Ganglioneuroma	25

Table 21   Principal calcitonin-secreting tumours

Tumour location	Tumour type	Orthoendocrine products	Paraendocrine products
Anterior pituitary	Usually adenoma	ACTH/MSH GH PRL, Gn	
Thyroid	Medullary carcinoma	CT	ACTH/MSH, prostaglandins, insulin, gastrin, SRIF
Adrenal gland	Phaeochromocytoma	Catecholamines	ACTH, FSH, HCG, VIP, insulin
Paraganglia	Chemodectoma, Ganglioneuroma, Neuroblastoma	Catecholamines	ACTH/MSH, CT, VIP
Thymus	Thymoma		CT, ACTH/MSH
Lung	Oat-cell carcinoma and carcinoids	5-HT, histamine	Most peptide hormones, e.g. ACTH/MSH, ADH, Gn, TSH, GHRH (GRF), SRIF, PRL, CT, PTH, insulin, oxytocin
Skin	Melanoma		Gastrin, ADH, FSH, PTH, PRL, ACTH
GI tract/pancreas	Carcinoids	5-HT, histamine	Most peptide hormones, e.g. ACTH/MSH, Gn, CT, PTH, VIP, enkephalins, endorphins, gastrin
Pancreatic islets	Gastrinoma ACTHoma Insulinoma Glucagonoma SRIFoma PPoma VIPoma	Insulin Glucagon SRIF PP VIP	Gastrin ACTH Secretin, GIP, neurotensin, Gn, ADH, oxytocin, CT, plus any of the other hormones typical of pancreatic islet tumours, (ortho- or paraendocrine)
Stomach/small intestine	Gastrinoma SRIFoma GHRHoma (GRFoma)	Gastrin SRIF	ACTH GHRH (GRF)

Table 22   A classification of APUDomas[119]

– Hypercalcaemia and primary
  hyperparathyroidism
Calcium, of course, is the principal stimulant of calcitonin secretion.

– Renal disorders
The general consensus appears to be that both acute
and chronic renal failure are associated with raised
basal levels of immunoreactive calcitonin[136–142], although some authors have reported depleted levels,
notably in early-stage disease and in juvenile renal failure[143]. Secretory and metabolic mechanisms are

thought to be implicated in the elevation of calcitonin
levels; in patients with chronic renal insufficiency, for
example, excessive calcitonin is secreted in response
to raised serum levels of calcium or gastrin[144], different forms of the hormone apparently responding to
the different stimuli. One of these forms is a non-
dialysable high-molecular-weight calcitonin with
little or no activity, while another is its biologically
active and dialysable monomer (Fig.29). The two
forms can be detected and distinguished by radioimmunoassay. The metabolic clearance of calcitonin is
also diminished in renal insufficiency[139,145,146].

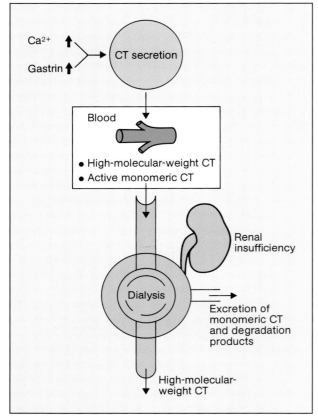

Fig. 29   Diagrammatic representation of the mechanism by
which the calcitonin level is raised in chronic renal insufficiency

– Pancreatitis

Findings in patients with pancreatitis are often inconsistent, with raised calcitonin levels in some and
normal levels in others[136,147–155]. Glucagon is claimed
by many authors to be responsible for stimulating calcitonin secretion[136,148,151-153], but others think that a
substance with calcitonin-stimulating activity may be
produced in the pancreas itself or in adjacent gastrointestinal cells[119,155]. Another possible explanation is
that the inflamed pancreas may secrete substances
that interfere with radioimmunoassay, leading to artefacts which produce "false-elevated" plasma levels
of immunoreactive calcitonin.

– Heroin addiction

Mean serum concentrations of calcitonin in a group of
heroin addicts were reported to be higher than in controls matched for age and sex (Fig. 30), even in the
absence of clear signs of impaired kidney or liver function and of abnormal serum calcium and phosphate
levels. This suggests that opioids stimulate the secretion of calcitonin directly or by some unknown
indirect mechanism[109].

– Other conditions

See Table 23.

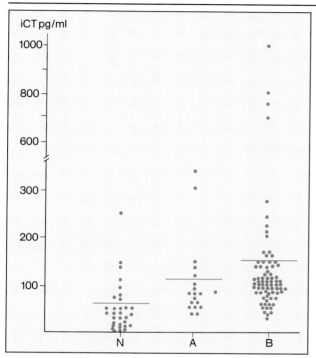

Fig. 30   Calcitonin levels in normal subjects (N) and in two groups of heroin addicts aged 17–35 years: A = addicts consuming < 500 mg/day (n = 19), B = addicts consuming ≥ 500 mg/day (n = 75)[109].

Conditions associated with calcitonin underproduction	
Osteoporosis	
– Postmenopausal/postovariectomy	114
– Senile	113
– Secondary to endocrine disorders	105,115
Non-goitrous congenital cretinism	116
Acromegaly (?)	105

Conditions associated with calcitonin overproduction	
Calcitonin-secreting tumours	
– Medullary carcinoma of the thyroid	110
– Other tumours (see Table 22)	
Hypercalcaemia	
Neonatal hypocalcaemia	132
Primary hyperparathyroidism (controversial)	133 (134)
Pseudohypoparathyroidism	135
Renal disorders	
– Renal failure (acute and chronic)	136–142
– Renal dialysis (esp. high-molecular-weight forms of calcitonin)	52
Pancreatitis	136, 147–155
Heroin addiction	109

Other conditions associated with raised calcitonin levels	
– Graves' disease	72
– Atrophic gastritis	72
– Acute gastritis	72
– Pernicious anaemia	72
– Peptic ulcer	72
– Gastrointestinal bleeding	72
– Stress	
– Thyroid surgery	156
– Hepatic surgery	157
– Toxic shock	158
– Myocardial infarction	159
– Difficult labour	160
– Trauma	161
– Lithium intoxication	107

Table 23   Pathological states associated with altered calcitonin secretion in man

Calcitonin levels in cerebrospinal fluid and milk

Immunoreactive calcitonin was found in the cerebrospinal fluid of 75% of 63 male subjects, with a mean level of 11.1 ± 1.3 pg/ml and a range of < 2–55 pg/ml (Table 18). The concentration ratio CT plasma : CT CSF was 2.3 : 1 (Fig. 31)[64].

In another study[162] the concentration of immunoreactive calcitonin in human milk was found to be 10–40 times higher than the serum level. In milk samples collected 1 week and then 3 months post partum, levels of 2.46 ± 0.99 ng/ml and 0.58 ± 0.08 ng/ml respectively were measured, the decline being quite pronounced within a few days of delivery (Fig. 32). The high concentration in milk compared with serum suggests the local production of iCT within the mammary gland or a specific transport and concentration mechanism.

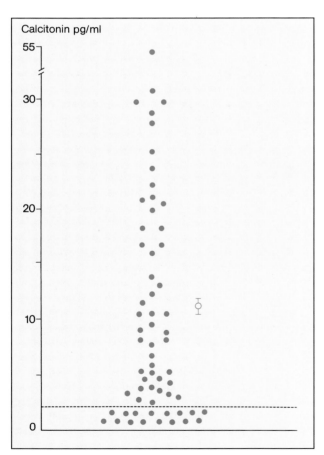

Fig. 31   Cerebrospinal fluid concentrations of calcitonin in 63 male subjects. The mean ± SEM for the whole population was 11.1 ± 1.3 pg/ml (⌀). The broken line represents the assay detection limit[64].

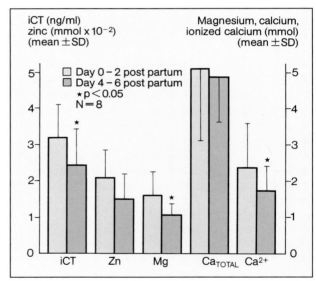

Fig. 32   Concentrations of immunoreactive calcitonin, zinc, magnesium, total calcium and ionized calcium in human milk from 8 mothers. Each mother gave two samples, one within two days post partum and the other 4–6 days after delivery[162].

The physiological functions of calcitonin

Endogenous calcitonin, with its variable and fluctuating basal level in healthy people, has a complex physiological role which is still not well understood. Indeed, it may have several roles (Fig. 33), notably in the regulation of functions involving calcium. This is particularly important since calcium, as one of the major constituents of the body's internal milieu, plays a vital role in structural (skeletal and muscular) and other systems (endocrine, nervous and circulatory). Calcium is involved, for example, in the control of cellular permeability, neuromuscular excitability, muscular contraction, the activation of certain enzymes (lipase, succinyl dehydrogenase, trypsinogen, ATPase), in endocrine secretion, cardiac function and blood coagulation (as reviewed in [92]).

The principal role of calcitonin is in the regulation of mineral (especially calcium) metabolism, chiefly in helping the body to deal with episodes of 'calcium stress', i.e. preventing calcium depletion or excess. It also, however, directly or indirectly controls the movement of other ions, such as phosphate and magnesium, as part of the process of maintaining ionic equilibrium. Calcitonin, which can thus be described as a hormone of mineral regulation, exerts an effect at many levels:

In bone

The principal structural features of bone are shown in Figure 34. In various stress situations (growth, pregnancy, lactation, after eating) calcitonin protects the skeleton by exerting a direct inhibitory effect on the osteoclasts involving both their activity and their number. In controlling osteoclast activity, it presumably also regulates osteoblast activity and, therefore, bone formation[163], although this aspect of its effect requires further elucidation.

Bone undergoes continual renewal within remodelling units which have been termed 'basic multicellular units'[164]. These units consist of two types of cells, osteoclasts and osteoblasts, which differ in origin (see Fig. 35), morphology and function, but are probably linked by a still unknown coupling factor.

Osteoclasts, which probably derive from the same committed progenitor cells as monocytes[47,48], measure 30–100 μm in diameter and normally have 6–20 nuclei and a basophil cytoplasm containing a well-developed Golgi apparatus. The surface of the cell membrane which is in contact with bone is deeply infolded in the same way as that of a macrophage; this is the highly characteristic brush border of the osteoclasts. Osteoclasts are constantly changing in size and shape and are highly mobile[49], moving along the bone surface in the performance of their basic role, the resorption of bone. The mechanism by which they achieve this is poorly understood; possibly it entails resorption of the bone matrix by a specific process involving its extracellular digestion by proteolytic enzymes or a collagenase along the brush border in an acid environment[239]. Their activity results in the excavation of pits known as resorption lacunae of Howship.

Osteoblasts, which are derived from committed progenitor cells probably common to fibroblasts[165–168], are 10–20 μm in diameter and ovoid or cylindrical in shape. They have a large nucleus at the pole furthest away from the bone surface, a perinuclear halo, and cytoplasm which is intensely basophilic owing to the presence of a highly developed endoplasmic reticulum. The osteoblasts (for properties, see Table 24) secrete the protein matrix of bone, which consists mainly of type I collagen, or tropocollagen, the basic unit of collagen fibre, but also contains mucopolysaccharides. They are separated from the zone of bone that is in course of formation by a layer of a pre-osseous substance called osteoid. The deep zone of this osteoid border, where inorganic salts are being deposited, is called the mineralization front.

Once its bone-forming activity is completed, the os-

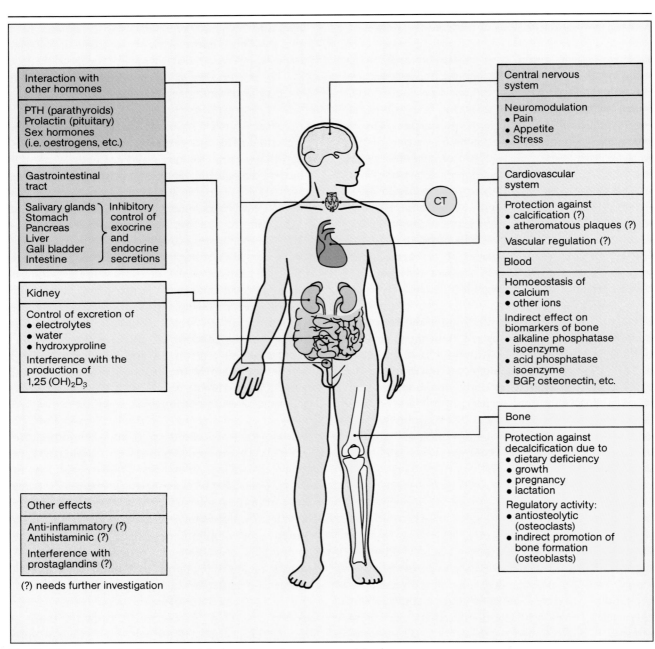

Interaction with other hormones

PTH (parathyroids)
Prolactin (pituitary)
Sex hormones
(i.e. oestrogens, etc.)

Gastrointestinal tract

Salivary glands ⎫
Stomach ⎪ Inhibitory
Pancreas ⎬ control of
Liver ⎪ exocrine
Gall bladder ⎪ and
Intestine ⎭ endocrine
secretions

Kidney

Control of excretion of
• electrolytes
• water
• hydroxyproline

Interference with the production of
$1,25 (OH)_2D_3$

Other effects

Anti-inflammatory (?)
Antihistaminic (?)

Interference with prostaglandins (?)

(?) needs further investigation

Central nervous system

Neuromodulation
• Pain
• Appetite
• Stress

Cardiovascular system

Protection against
• calcification (?)
• atheromatous plaques (?)

Vascular regulation (?)

Blood

Homoeostasis of
• calcium
• other ions

Indirect effect on biomarkers of bone
• alkaline phosphatase isoenzyme
• acid phosphatase isoenzyme
• BGP, osteonectin, etc.

Bone

Protection against decalcification due to
• dietary deficiency
• growth
• pregnancy
• lactation

Regulatory activity:
• antiosteolytic (osteoclasts)
• indirect promotion of bone formation (osteoblasts)

CT

Fig. 33   Diagram showing the main physiological actions of endogenous calcitonin

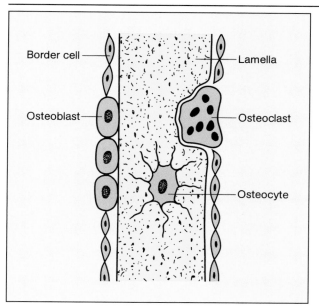

Fig. 34a   The various types of bone cells

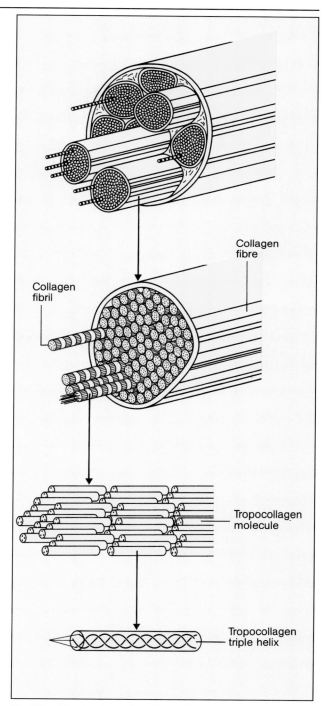

Fig. 34b   The structure of a collagen fibre

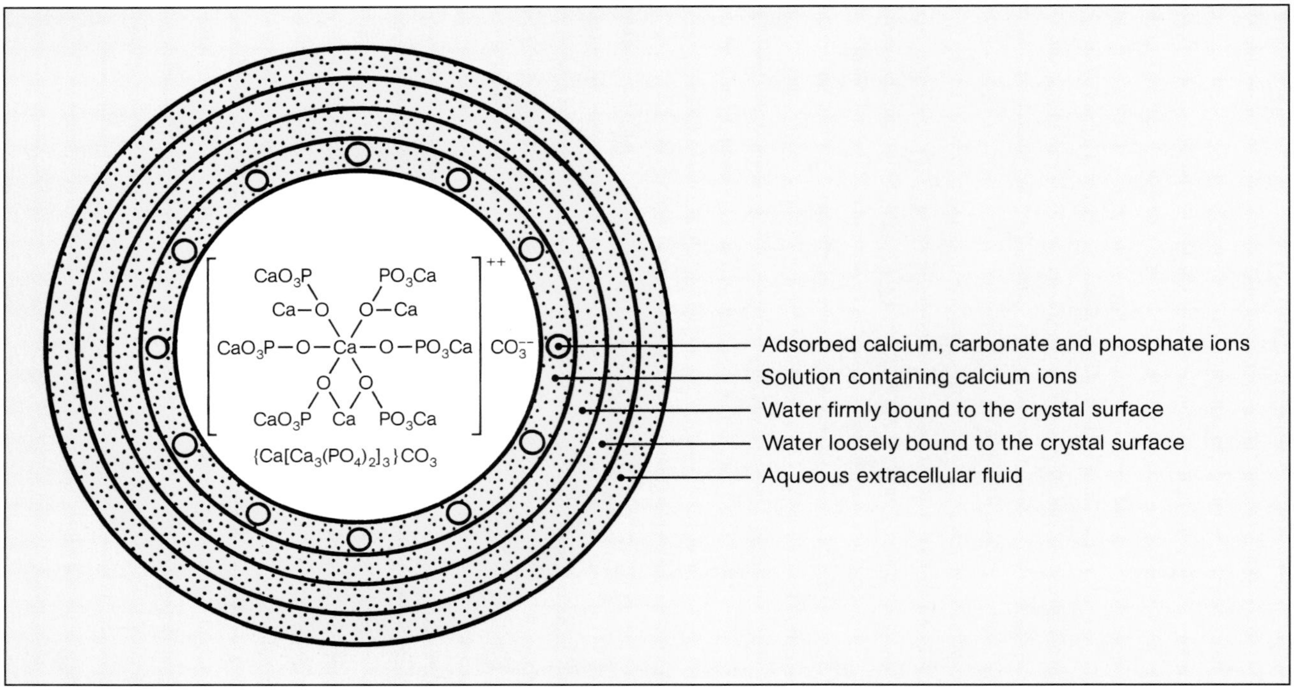

Fig. 34c   The structure of hydroxyapatite

Collagen synthesis, type I
Osteocalcin synthesis (BGP)
Osteonectin synthesis
Collagenase synthesis
Alkaline phosphatase synthesis
Hormone responsiveness
         PTH
         $1,25 \, (OH)_2 \, D_3$

Practical definition of osteoblast cells:

Cells capable of producing a type−I collagen matrix,
which is subsequently mineralized.

Table 24   Properties of osteoblasts (E. Schmid, personal
communication)

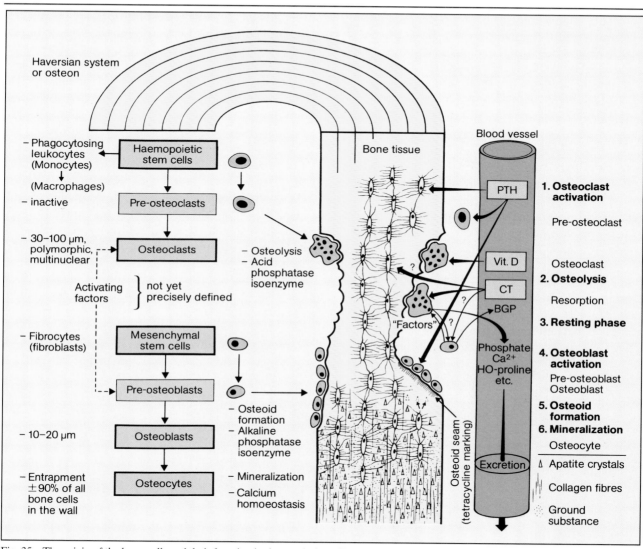

Fig. 35   The origin of the bone cells and their function in the regulation of bone metabolism

teoblast becomes entrapped within the mineralized bone and is transformed into an *osteocyte*, an ovoid or spindle-shaped cell with numerous slender processes. The osteocytes make up approximately 90% of the cells of bone; their total surface area plus that of the lacunae which they occupy is of the order of 1200 m^2, or 100 times greater than that of the trabeculae. This large surface area greatly facilitates exchange between the osteocytes and the neighbouring tissues in their (probably) important role in the control of blood calcium level and the mineralization of bone tissue.

The factors involved in the bone formation/resorption cycle are still not well understood, but include the following (see Table 25):

Parathyroid hormone (PTH), a polypeptide containing 84 aminoacid units which is synthesized in special secretory cells of the parathyroid. PTH stimulates osteoclast activity indirectly by increasing their number, activity and life span, thus exerting a potent, if indi-

rect, hypercalcaemic effect and encouraging bone resorption. It can also directly stimulate the proliferation or, equally, inhibit the activity of osteoblasts (which appear to possess PTH receptors[169]), depending on its concentration. Recently it has been shown *in vitro* that, in the presence of PTH, osteoblast-like cells release a soluble factor that stimulates osteoclastic bone resorption[169].

Calcitonin, which reduces the activity of osteoclasts (e.g. their motility) and their rate of proliferation, thus slowing bone resorption and lowering plasma calcium[6,10,11,170,171]. It is also possible that it exerts an effect on osteoblasts, either indirectly via the osteoclasts or by stimulating bone formation[163,172,173]. Recent findings suggest that a direct effect of calcitonin on osteoblasts can be demonstrated *in vitro* (L.J. Deftos; D.J. Baylink and J.R. Farley, personal communications).

Vitamin D, which acts on bone tissue through several of its metabolites, the most important of which is 1,25(OH)$_2$D$_3$. This stimulates the mobilization of bone calcium by increasing the intestinal absorption of calcium as required, and is also involved in the maintenance of calcium balance.

A number of other factors besides calcitonin are also thought to be involved in maintaining the integrity of bone:

Oestrogens, oestrogen replacement therapy (with and without progestagens) being well established in prevention of the onset of postmenopausal osteoporosis, although it is not yet clear whether these hormones have a direct effect on bone cells and what their real adverse effects are.

Corticosteroids, which reduce osteoblast activity and thus the formation of bone.

Thyroid hormone, which appears to stimulate both the osteoblasts and the osteoclasts.

Agent	Effect on bone resorption		Effect on bone formation	
	Direct	Indirect	Direct	Indirect
Calcium regulatory hormones				
Parathyroid hormone	↑	↑	↓	↑
1,25(OH)$_2$D$_3$	↑	↓	↓	?
Calcitonin	↓	–	–	–
Systemic hormones				
Glucocorticoids	↓	↑	↓	↓
Insulin	–	–	↑	↑
Thyroxin	↑	–	?	↑
Sex hormones	–	↓	–	↑
Growth hormone	–	–	–	↑
Cytokines and related factors see Table 26				

Table 25   Hormonal regulation of bone and cartilage metabolism. Agents which have been tested for their direct effects *in vitro* are listed as increasing (↑), decreasing (↓) or not changing (−) bone resorption or formation. Where there is evidence for an important indirect effect, mediated through another system or local factor, the dominant direction is indicated.

Cytokines and related factors of skeletal (bone and cartilage) metabolism. An increasing number of factors (known as cytokines and chondrokines) involved in bone formation and/or resorption are being isolated and characterized (Tables 25, 26 and 27). However, it is not yet clear whether they act on or are products (markers) of these processes. Also, it seems probable that some of them might well be derived from the same substance through modification during extraction. Indeed, they may *be* the same substance, their apparent diversity merely reflecting poor specificity of the analytical methods used. Some cytokines also appear to be potential bone-resorbing tumour products (Table 28).

Factor	Type of malignancy
Lymphotoxin (or tissue-necrosis factor$_\beta$)	Myeloma Monocytic malignancy
Interleukin$_1$	Adult T-cell lymphoma
1,25 (OH)$_2$D$_3$	Breast cancer
Prostaglandins	Solid tumours
PTH-like factors	Several tumour types
Transforming growth factors	Many different types
Platelet-derived growth factor	

Table 28   Potentially bone-resorbing tumour products (U. Trechsel, personal communication)

Formation	Resorption
Chondrocyte-induced factor (= transforming growth factor$_2$?)	Interleukin$_1$ (= osteoclast-activating factor?)
Osteogenin	Interleukin$_2$
Skeletal growth factors	Interleukin$_3$ (= colony-stimulating factor?)
Insulin-like growth factors	
Transforming growth factors	
Bone-derived growth factor	Macrophage-derived growth factor
Epidermal growth factor	
Bone morphogenetic protein	Prostaglandins
Fibroblast growth factors	Lymphotoxins (= tumour-necrosis factors?)
Platelet-derived growth factor	
Somatomedins (= insulin-like growth factors)	Interferon γ
Prostaglandin E$_2$	
Cartilage-derived factor	
Cartilage-derived growth factor	
Macrophage-derived growth factor	
Chondrogenic stimulating activator	
Mesenchymal-cell chemotactic activator	

Table 26   Cytokines and related factors involved in bone and cartilage metabolism

Bone formation and resorption are thus interdependent and closely linked both in time and location. This is borne out by the observation in compact bone that the osteons, or Haversian systems, are separated by irregular boundaries, or cement lines, a morphology which develops as the resorption of one Haversian system is followed by the birth of a new system. Thus in the bone remodelling unit, osteoclasts removing existing bone operate side by side with osteoblasts synthesizing new tissue to fill the resultant lacunae.

A	Probable factors	B	Possible factors
	1. Prostaglandins		1. Bone-derived resorption stimulator
	2. Insulin-like growth factors		
	3. Bone-derived growth factor		2. Bone-derived parathyroid hormone inhibitor
	4. Bone GLA protein (osteocalcin)		
	5. Osteonectin		3. Stimulator of prostaglandin release
	6. Collagen pro-peptides		
	7. Transforming growth factors		

Table 27   Biologically active factors of cultured rat bone (L.G. Raisz, personal communication)

The process starts, after a resting (quiescent) phase, with a phase of activation lasting from several hours up to 3 days. The mechanism of this activation is still unknown, but it results in the appearance of a focus of osteoclast activity to initiate the resorptive phase, which lasts 2–3 days. This is followed, after a 'reversal phase' of 3–10 days' duration, by the arrival of osteoblasts to begin the formation phase. This in turn lasts 3–4 days, and its completion is followed by a return to quiescence, which can last for anything from 1 week to 5 years (Fig. 36). This sequence of resorption (R) and formation (F), in which the R phase (the life span of a resorption lacuna) is shorter than the F phase, constitutes the temporal link between osteoclast and osteoblast activity. Physiologically, the level of bone remodelling activity, in which large numbers of remodelling units are involved, varies as requirements change. Many other factors besides calcitonin thus play a part in bone metabolism, and any disruption due to an excess or deficiency of any of these factors will give rise to a pathological process.

It has been postulated that the role of calcitonin in the control of osteoclast activity is to regulate, together with PTH, the movement of calcium between the extracellular, intracellular and mitochondrial compartments. Inside the cell, calcium may remain in the free state in the cytosol or be reversibly deposited in the mitochondria[175–177]. Calcitonin is claimed to promote this deposition, an effect which is increased by phosphates, which enhance the ability of the mitochondria to accumulate calcium[178,179]. PTH, on the other hand, is reported to indirectly promote calcium efflux from the mitochondria into the cytosol and thence in the direction of the extracellular fluid[180–182]. This active redistribution of calcium during osteolysis is accompanied by increased release of calcium from bone and by increased reabsorption by the renal tubules, while its net effect in terms of the gain or loss in total cell calcium, depends on the concentration ratio of intracellular to extracellular calcium (ionic activity)[178,180–183].

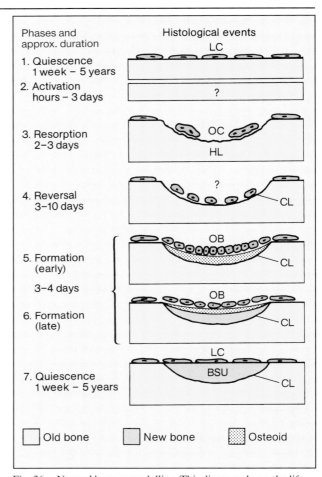

Fig. 36   Normal bone remodelling. This diagram shows the life history of a single bone remodelling unit, with construction of a single new bone structural unit (BSU). 1) Quiescent surface covered by flat lining cells (LC). 2) Resorbing surface with osteoclasts (OC) lying in a Howship's lacuna (HL). 3) Reversal phase, during which mononuclear cells of uncertain origin smooth over the resorbed surface and deposit the cement substance, seen as the cement line (CL) in histological sections. 4) Early-forming surface with young osteoblasts (OB) which have deposited a thick layer of osteoid on the cement line; mineralization has not yet begun. 5) Late-forming surface with old osteoblasts which have almost finished making osteoid; there is now a layer of new mineralized bone between the cement line and the osteoid seam. 6) Restoration of the quiescent surface with completion of the cycle of remodelling. The cavity has been completely refilled to form a new bone structural unit. The coordinated team of precursor cells, osteoclasts, reversal cells, and osteoblasts appearing sequentially in the same location to build a single bone structural unit comprise a single bone remodelling unit[174].

Others disagree with this hypothesis, reporting that in murine osteoclasts salmon calcitonin mobilizes calcium out of the mitochondrial compartment into two more superficial and rapidly exchanging compartments, without affecting the total content of the cell (J. Rosen, personal communication), while phosphate causes a marked rise in mitochondrial calcium, presumably via the deposition of calcium phosphate salts in mitochondrial matrix[178,179]. PTH-like agents (phosphodiesterase inhibitors) bring about a redistribution of intracellular calcium from the two rapidly exchanging pools into the mitochondrial compartment, i.e. they cause an increase in mitochondrial calcium (Fig. 37).

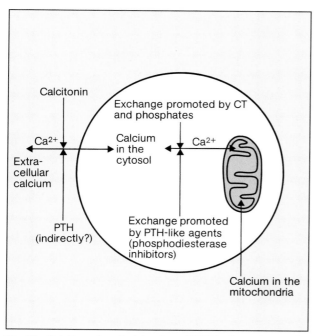

Fig. 37 Putative roles of calcitonin and PTH in the exchange of calcium between the extracellular and intracellular compartments and between the cytosol and the mitochondria of bone cells

Some authors suggest that the primary effect of calcitonin on bone mineralization is to regulate the uptake of phosphate by bone cells, the combination of phosphate and calcium leading to the precipitation of

hydroxyapatite. According to this hypothesis, the effect of calcitonin on blood calcium is simply a consequence of mineral nucleation after phosphate uptake[177,184] and is either distinct from its effect on bone resorption or else linked in some way via the inhibitory action of phosphate on resorption[185].

In the kidney

Although specific calcitonin receptors distinct from PTH and vasopressin receptors have been shown to be present in the kidney[39,40,186,187], calcitonin appears to play a minor, though direct, role in normal renal function, being involved to some extent in electrolyte and water excretion, and consequently in calcium homoeostasis. The cortex of the human kidney contains a calcitonin-linked adenylate cyclase, guanylate cyclase and Ca^{2+}-Mg^{2+}-dependent ATPase. Calcitonin also enhances 1α-hydroxylase activity in the proximal straight tubule, which increases the production of $1,25(OH)_2D_3$ from its substrate[188,189].

In calcium homoeostasis

The function of all the movements of calcium described above – especially those affecting bone and renal function – in which calcitonin and the other regulatory factors are involved, is to maintain a normal blood level of calcium during episodes of physiological hypocalcaemic or hypercalcaemic 'stress', e.g. due to eating (Fig. 38). The importance of endogenous calcitonin in this process has long been clear from the fact that removal of the C cells by thyroidectomy impairs the homoeostatic response to calcium-, PTH- or vitamin-D-induced hypercalcaemia, whereas if the thyroid is intact calcitonin secretion automatically increases in response to such hypercalcaemic stimuli so as to maintain calcium balance[176,177].

Calcium is the fifth most abundant element in the body, most of it being stored in the bones. It is also

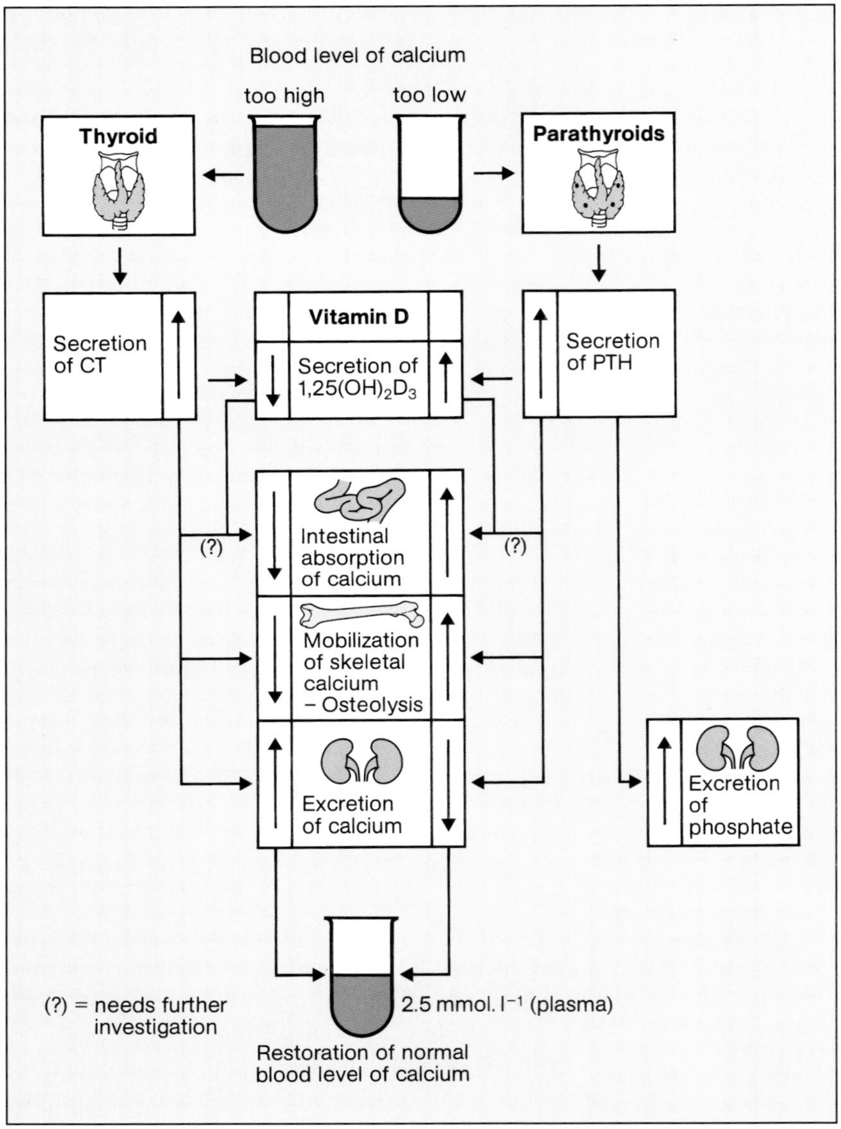

Fig. 38   The principal mechanisms regulating calcium homoeostasis

present in small quantities in the extracellular fluid and in soft-tissue cells. The normal human daily calcium requirement ranges from 200 to 2500 mg and the serum level is maintained at about 2.5 mmol/l [=10 mg/dl] (Fig. 39). Forty percent of this is bound to proteins (albumin), 10% is diffusible but complexed with anions such as citrate, phosphate and bicarbonate, and the remainder is in diffusible ionized form (Fig. 40). This is its physiologically most active form (Fig. 41) and the one which should therefore be assayed before the (diagnostic) significance of a given plasma calcium value can safely be interpreted.

Although in animals (and probably in man, too) some calcium absorption takes place throughout the gut – including the colon, when it progresses so far in an absorbable state[190-192] – the site of most extensive absorption is the more proximal segments of the small in-

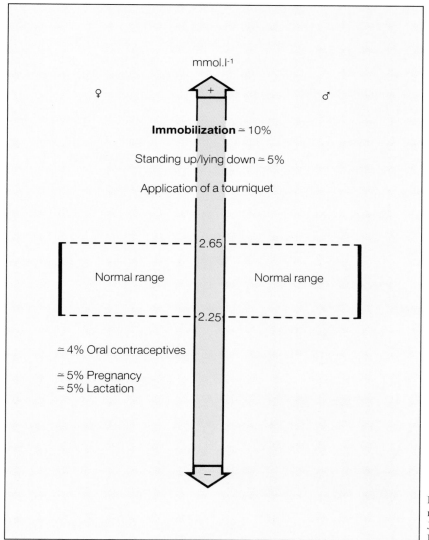

Fig. 39   Serum calcium: normal levels and minor, non-disease-related variations (G. Siest, J. Henny – by courtesy of Sandoz S.A., Rueil-Malmaison, France)

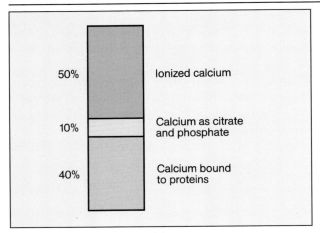

Fig. 40   Fractions of plasma calcium

testine. Approximately one-third of the amount ingested is actually absorbed. Absorption is thought to be increased by vitamin-D and PTH-induced $1,25(OH)_2D_3$ but reduced by glucocorticoids and complexes of insoluble salts. Calcium is eliminated in saliva, bile and pancreatic juice. It is also lost in sweat and in milk during lactation. Urinary excretion represents the net difference between the quantity filtered by the kidneys and the quantity reabsorbed (Fig. 42). Renal reabsorption is stimulated by PTH via an effect on the distal tubule, and by vitamin D via an effect on the proximal tubule.

● Abnormalities of calcium metabolism

Total calcium blood levels much below 2.25 or above 2.65 mmol/l (=9–11 mg/100 ml) indicate an abnormality of calcium metabolism (Fig. 43). This range is slightly lower in new-born infants[193] and slightly higher in children[194]; it varies with the plasma protein concentration.

Hypocalcaemia
may be due to calcium or vitamin-D deprivation, to lack of exposure to sunlight, to malabsorption, or to resistance to vitamin-D activity[136].

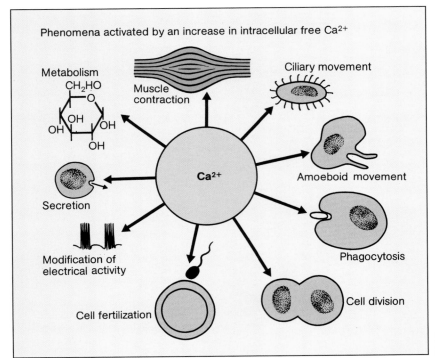

Fig. 41   Phenomena activated by an increase in intracellular free Ca^{2+}

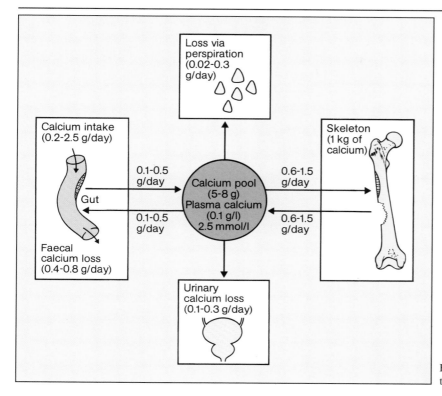

Fig. 42   Calcium balance and movements in the normal adult

When associated with malabsorption, hypocalcaemia is accompanied by inadequate blood levels of phosphorus, total protein and magnesium. Hypoparathyroidism may also be associated with hypocalcaemia, but in this case phosphataemia is raised rather than depressed. Neonatal tetany is due to hypocalcaemia, early neonatal hypocalcaemia probably being due to temporary hypoparathyroidism occurring in the offspring of mothers with hyperparathyroidism[193]. Neonatal hypocalcaemia may also supervene after hypernatraemia and acute infections. Hypocalcaemia is also frequently associated with advanced renal insufficiency accompanied by hyperphosphataemia, to the ingestion of excessive amounts of sodium fluoride, which forms an insoluble salt with calcium, and to massive transfusion of citrated blood (as reviewed in [195]). Hypocalcaemia also occurs in elderly people due to a combination of factors, as illustrated in Figure 44.

Hypercalcaemia,
for which a number of factors are responsible (Tables 29 and 30), is commonly associated with hyperparathyroidism, raised blood levels of $1,25(OH)_2D_3$, hypophosphataemia and/or malignancy. A serum calcium level above 3.25 mmol/l($=13$ mg/100 ml) may cause the following:

– Renal lesions affecting the glomeruli and the tubules, possibly progressing to severe renal failure through massive deposition of calcium salts in the kidney

– Mental disturbances, confusion, lethargy

– Arrhythmias, formation of calcareous aggregates in the soft tissues, reduced response to analgesics

– Syncope and death

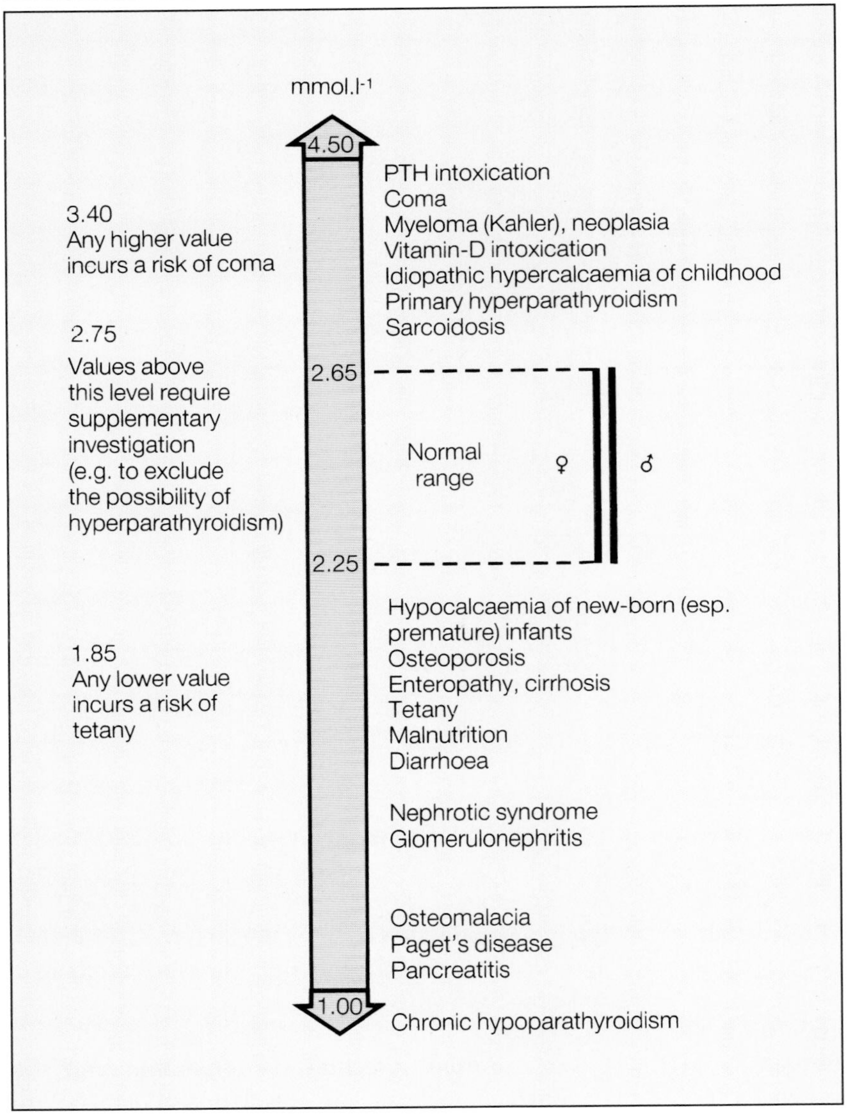

mmol.l^{-1}

4.50

PTH intoxication
Coma
Myeloma (Kahler), neoplasia
Vitamin-D intoxication
Idiopathic hypercalcaemia of childhood
Primary hyperparathyroidism
Sarcoidosis

3.40
Any higher value
incurs a risk of coma

2.75
Values above
this level require
supplementary
investigation
(e.g. to exclude
the possibility of
hyperparathyroidism)

2.65

Normal
range   ♀   ♂

2.25

Hypocalcaemia of new-born (esp.
premature) infants
Osteoporosis
Enteropathy, cirrhosis
Tetany
Malnutrition
Diarrhoea

1.85
Any lower value
incurs a risk of
tetany

Nephrotic syndrome
Glomerulonephritis

Osteomalacia
Paget's disease
Pancreatitis

1.00   Chronic hypoparathyroidism

Fig. 43   Serum calcium: pathological
alterations (G. Siest, J. Henny – by courtesy of
Sandoz S.A., Rueil-Malmaison, France)

Artefactual

> Hyperproteinaemia
>> Venous stasis during blood collection
>> Hyperalbuminaemia (e.g. due to hyperalimentation)
>> Hypergammaglobulinaemia (e.g. due to a myeloma)

Malignancy

> Solid tumours
> Haematological
>> Myeloma
>> Lymphoma
>> Leukaemia

Endocrinological

> Primary hyperparathyroidism
> Multiple endocrine adenomatoses, types I and II
> Inappropriate secondary hyperparathyroidism (renal failure)
> Hyperthyroidism
> Hypoadrenalism
> Adrenocortical deficiency states (e.g. Addison's disease)

Drugs

> Vitamin D intoxication
> Thiazides
> Calcium
>> Milk alkali syndrome
>> Dialysis

Granulomatous Disorders

> Sarcoidosis
> Tuberculosis
> Berylliosis

Paediatric Disorders

> Infantile hypercalcaemia
> Hypophosphataemia

Immobilization

> Paget's disease
> Young patients

Table 29   Causes of hypercalcaemia

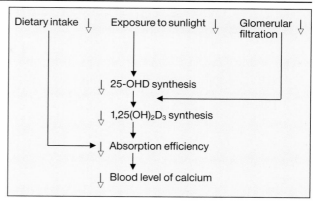

Fig. 44   Mechanism of hypocalcaemia in old age

Condition	Hypercalcaemic factors
Primary hyperparathyroidism	PTH $+ 1,25\,(OH)_2D_3$
Malignant tumours	PTH-like factors
	Transforming growth factors
	Tissue-necrosis factors
	(lymphotoxins)
	Interleukins
	$1,25\,(OH)_2D_3$
	Colony-stimulating factors
	Prostaglandins
Sarcoidosis	$1,25\,(OH)_2D_3$
Vitamin-D intoxication	$25\text{-}OHD_3$
	$1,25\,(OH)_2D_3$
Familial hypercalcaemia	?

Table 30   Factors giving rise to hypercalcaemia

Diurnal variations in blood calcium levels

In healthy post-menopausal women, blood levels of ionized calcium – and phosphorus – are reported[196] to be subject to diurnal variations, with a characteristic peak, trough and periodicity (peaks in late evening and early morning). This pattern differs from that previously reported[240] for young men (peak in mid-morning), suggesting that mineral homoeostasis may be altered in post-menopausal women. On the other hand, there is no unanimity among authors that blood calcium is subject to diurnal variation.

In the gastrointestinal tract

Endogenous calcitonin seems to affect the functioning of both endocrine and exocrine glands and glandular cells of the gastrointestinal tract under physiological conditions and, probably, as a function of food ingestion[197]. This conclusion is based on the following clinical and experimental evidence:

– The C cells and the secretory cells of the gastrointestinal tract have a common origin[120,198,199].

– The disturbance of calcium metabolism which occurs in primary hyperparathyroidism is often accompanied by the triad of peptic ulcer, pancreatitis and cholelithiasis[200,201].

– Induced hypercalcaemia increases the acidity of the gastric juice in normal subjects[200,201].

– Several hormones of the gastrointestinal tract are potent calcitonin secretagogues (as reviewed in [52]; Table 19).

– The presence of calcium ions is essential for the release of insulin[202].

It seems, therefore, that calcitonin is not merely a hormone with effects on calcium metabolism and bone turnover. It also has definite effects on the gastrointestinal tract, secretions of which are dependent on the calcium concentration, and therefore on the activity of calcitonin.

In the central nervous and endocrine systems

Calcitonin probably interacts with many other hormones besides those of the gastrointestinal tract. Calcitonin levels, for example, are affected by levels of growth hormone and prolactin. Calcitonin, or a related immunoreactive compound, and specific receptors for calcitonin have been shown to be present in the central nervous system in zones involved in the control of pain perception, appetite, prolactin secretion, the initiation of lactation[22,24,36,37,203,204] and protection against stress. Both the hormone and its binding sites have been found in the hypothalamus, the control centre of the autonomic nervous system, suggesting a neuromodulator role for calcitonin (as reviewed in [205]), although as yet no hard evidence is available.

Other physiological roles

Many other physiological roles have been claimed for calcitonin, including anti-inflammatory activity, prevention of calcium deposition in the cardiovascular system (and even of the formation of atheromatous plaques), an antihistaminic effect, interaction with prostaglandins, and an anti-stress effect (as reviewed in [92]). Most of these effects, however, are speculations based on experimental work and, while they might occur with pharmacological concentrations of the hormone, they are minimal at physiological blood levels.

Thus endogenous calcitonin acts mainly on calcium, but also on other electrolytes such as phosphates[177]. It protects the skeleton by inhibiting osteoclast activity and thereby reducing bone resorption and remodelling. Its indirect action on the kidney and the gastrointestinal tract facilitates the regulation of extracellular calcium. However, its wide distribution throughout the body, including the central nervous system, and its presence in animals that have no bony skeleton suggest that calcitonin may possess other properties in addition to its action on bone. According to Deftos and Parthemore[206] and to Copp[207], the common denominator in the various physiological actions of calcitonin could well be the modulation of calcium flux across the membranes of a number of different types of cells, and thus of the intracellular-extracellular distribution of calcium in various systems. At CNS level, for example, control of pain by calcitonin might be explained by changes in calcium flux between nerve tissue and the cerebrospinal fluid.

Biochemical markers of calcitonin activity and bone metabolism

Calcitonin modulates the activity of various enzyme systems, including the adenylate-cyclase/phosphodiesterase system in connection with the formation of cyclic AMP and activation of glucose-6-phosphate dehydrogenase, 5'-nucleotidase, alkaline phosphatase, acid phosphatase, Ca^{2+}-Mg^{2+}-dependent ATPase and Na^+-K^+-dependent ATPase[31,187,208,209]. Moreover, it probably exercises these functions in other cells in addition to its primary target (the bone cells), notably in chondrocytes[210], embryonic periosteal cells[211], kidney cells[212], liver cells[213–216], thymocytes[217,218] and human leukaemic lymphocytes[43]. As a result, these and other biochemical factors associated with bone turnover and with calcitonin activity may be used as markers in physiological, pathological (e.g. Paget's disease[219], Fig. 45) or therapeutic situations involving the hormone. The principal of these markers are:

Cyclic AMP

As we have seen, calcitonin binds specifically to receptors linked to adenylate cyclase located on the membranes of its target tissue (principally bone and kidney) cells, triggering the release of cyclic AMP by the cell with a consequent rise of levels in blood and, to a lesser extent, urine[31,35]. This cyclic AMP, which is thus different in origin and less in quantity compared with that formed at various receptors in response to PTH[31,220], is thought to be calcitonin's second messenger[221]. The action of calcitonin can be mimicked by substances such as dibutyryl-cyclic AMP[222]. However, this rise in cyclic AMP seems not to be specific to calcitonin and may be partly due to PTH.

Alkaline phosphatase (Figs 46–49)

Assay of serum skeletal alkaline phosphatase often provides a valuable indication of the extent of bone remodelling activity. Levels are particularly high in Paget's disease, when the osteoblasts are more active, probably as a result of increased bone formation.

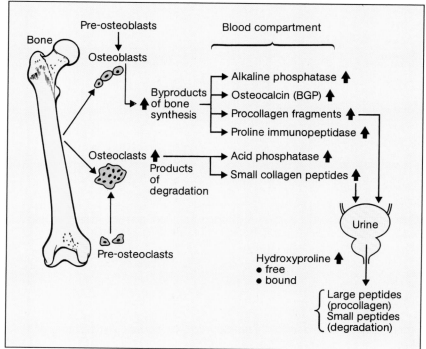

Fig. 45   Principal biochemical markers of Paget's disease

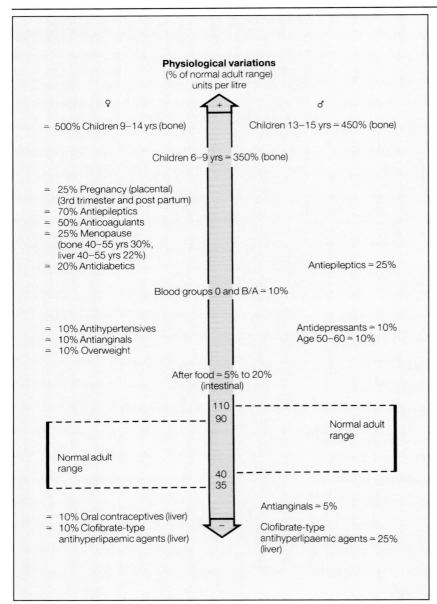

Fig. 46   Total alkaline phosphatase (serum): normal physiological and iatrogenic variations (G. Siest, J. Henny – by courtesy of Sandoz S.A., Rueil-Malmaison, France)

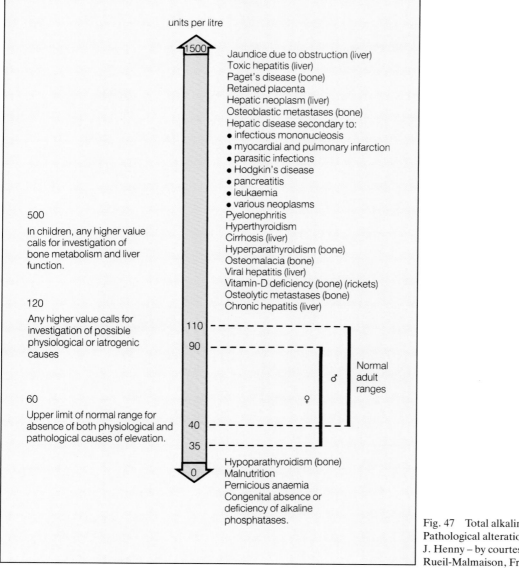

units per litre

1500 — Jaundice due to obstruction (liver)
Toxic hepatitis (liver)
Paget's disease (bone)
Retained placenta
Hepatic neoplasm (liver)
Osteoblastic metastases (bone)
Hepatic disease secondary to:
- infectious mononucleosis
- myocardial and pulmonary infarction
- parasitic infections
- Hodgkin's disease
- pancreatitis
- leukaemia
- various neoplasms
Pyelonephritis
Hyperthyroidism
Cirrhosis (liver)
Hyperparathyroidism (bone)
Osteomalacia (bone)
Viral hepatitis (liver)
Vitamin-D deficiency (bone) (rickets)
Osteolytic metastases (bone)
Chronic hepatitis (liver)

500

In children, any higher value
calls for investigation of
bone metabolism and liver
function.

120

Any higher value calls for
investigation of possible
physiological or iatrogenic
causes

60

Upper limit of normal range for
absence of both physiological and
pathological causes of elevation.

110
90

♂
♀

Normal
adult
ranges

40
35

0 — Hypoparathyroidism (bone)
Malnutrition
Pernicious anaemia
Congenital absence or
deficiency of alkaline
phosphatases.

Fig. 47   Total alkaline phosphatase (serum):
Pathological alterations (G. Siest,
J. Henny – by courtesy of Sandoz S.A.,
Rueil-Malmaison, France)

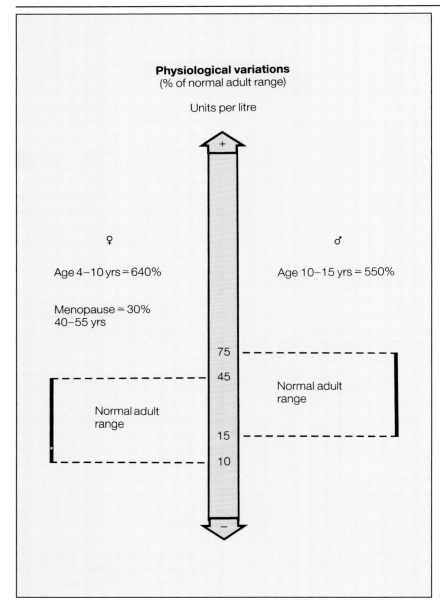

Physiological variations
(% of normal adult range)

Units per litre

♀

Age 4–10 yrs ≃ 640%

Menopause ≃ 30%
40–55 yrs

♂

Age 10–15 yrs ≃ 550%

75

45   Normal adult
      range

Normal adult
range

15

10

Fig. 48   Normal physiological variations in
serum levels of skeleton-derived alkaline
phosphatase (G. Siest, J. Henny – by courtesy
of Sandoz S.A., Rueil-Malmaison, France)

Acid phosphatase

This is thought to be the enzyme which is active in osteoclasts. Therefore, serum tartrate-acid-resistant
phosphatase might reflect bone resorption.

Hydroxyproline

Hydroxyproline is a non-essential aminoacid formed
by hydroxylation of proline residues linked to poly-
somes during collagen synthesis, the process being

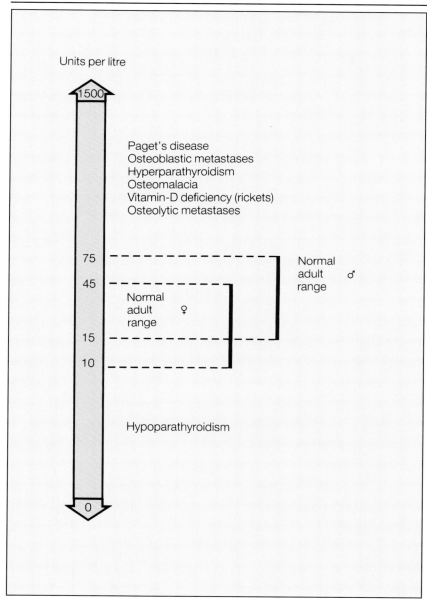

Units per litre

1500

Paget's disease
Osteoblastic metastases
Hyperparathyroidism
Osteomalacia
Vitamin-D deficiency (rickets)
Osteolytic metastases

75

45

Normal
adult   ♂
range

Normal
adult   ♀
range

15

10

Hypoparathyroidism

0

Fig. 49   Pathological changes in serum levels
of skeleton-derived alkaline phosphatase
(G. Siest, J. Henny – by courtesy of Sandoz
S.A., Rueil-Malmaison, France)

catalysed by proline hydroxylase. It is found in blood and urine either as the free aminoacid or linked to peptides (hypropeptides). It is broken down in the liver and kidney. Renal clearance is slow and there is significant reabsorption via the tubules. In bone dis-eases characterized by a high level of collagen turn-over, changes in blood and urinary hydroxyproline levels can be used as biochemical indicators of the ex-tent of collagenolytic activity and thus of bone resorp-tion[219].

Total urinary hydroxyproline is particularly useful for diagnosis and the monitoring of treatment[223].

Urine for hydroxyproline assay must be collected in fractions, starting not less than 3 days after institution of a low-collagen diet. Dialysable hydroxyproline can be assayed as a measure of bone resorption and non-dialysable hydroxyproline as a measure of bone formation[219,223,224].

Aminoacid residues	49
3 GLA and 1 S.S/molecule	
Molecular weight	5800
Solubility (mg protein/ml)	> 100
Precursor	9000 Mr (intracellular only)
Occurrence	Bone, dentine and serum
Synthesis	Osteoblasts and dentoblasts
Extraction from bone	By demineralization
GLA-dependent	Binds 3 Ca^{2+}/molecule
Properties	Binds strongly to hydroxyapatite
Regulation	Synthesis stimulated by 1,25(OH)$_2$D$_3$

Table 31   Principal characteristics of bone GLA protein (BGP)[231]

Bone GLA protein

Bone GLA protein (BGP), or osteocalcin, is the most abundant of the non-collagen proteins of bone (approximately 25% in adults). Composed of a single chain of 49 aminoacids, it is a specific, vitamin-K-dependent protein containing residues of γ-carboxyglutamic acid, which binds to both calcium and hydroxyapatite (Table 31). BGP also circulates in the blood, where it can be measured by radioimmuno-assay[225-227]. Serum levels in normal adults range from about 5 to 10 ng/ml (mean = 7 ng/ml) and increase with age (Table 32), especially in women. Levels are also higher in children than in adults, though these differences are less pronounced if values are corrected for creatinine clearance, which decreases with age. Serum BGP correlated positively with age (r = 0.32, p < 0.001) after creatinine clearance was fixed, but not with creatinine clearance when age was fixed (r = 0.04, n.s.)[225] (Table 33). Levels are higher than normal in patients with metabolic bone diseases characterized by increased turnover (e.g. Paget's dis-

Age range and mean age per decade	No. of subjects	Serum BGP	Serum alkaline phosphatase	Urinary hydroxyproline	Serum iPTH	Urinary cAMP	24–h creatinine clearance
yr		ng/ml	U/litre	mg/dl GF	µEq/ml	mmol/dl GF	ml/min
30–39 (35)	16	4.4 ± 0.4	15 ± 1.0	24 ± 2.3	44 ± 4.9	2.68 ± 0.10	99 ± 5.0
40–49 (45)	12	5.3 ± 0.5	21 ± 2.1	23 ± 3.9	81 ± 10.8	2.71 ± 0.21	93 ± 7.4
50–59 (55)	45	6.5 ± 0.4	27 ± 1.1	31 ± 1.4	60 ± 3.6	3.45 ± 0.16	90 ± 3.1
60–69 (65)	35	7.0 ± 0.2	26 ± 0.6	31 ± 0.9	80 ± 3.3	3.48 ± 0.08	82 ± 2.7
70–79 (74)	19	7.6 ± 0.6	29 ± 1.3	29 ± 2.0	96 ± 10.5	3.87 ± 0.19	68 ± 3.5
80–89 (85)	39	8.2 ± 0.4	26 ± 1.2	36 ± 2.3	99 ± 6.9	3.99 ± 0.17	52 ± 2.9
90–94 (92)	8	8.9 ± 0.9	28 ± 2.3	36 ± 3.2	101 ± 12.5	3.69 ± 0.16	37 ± 7.3

Table 32   Mean values (± SEM) per decade for serum BGP and other relevant biochemical measurements in 174 women[225]

ease), and lower than normal in patients with low bone turnover (Table 34).

Serum BGP originates from new cellular synthesis, probably by osteoblasts, and might reflect bone turnover rate in general or, more specifically, bone formation. It seems to be a more sensitive indicator of bone disease than alkaline phosphatase and levels appear to be lowered by calcitonin[227-229] and raised by PTH and 1,25(OH)$_2$D$_3$. Warfarin decreases serum levels[230], producing a non-γ-carboxylate BGP which cannot bind hydroxyapatite. An antiserum capable of detecting the complete molecule (all 49 aminoacids) can reportedly (L.J. Deftos, personal communication) be used to assess bone formation and an antiserum for BGP fragments to assess the level of resorption. Measurement of serum BGP might thus be a useful diagnostic aid in metabolic bone diseases and in assessing the effects of specific treatments ([225-227,229-232] and L.J. Deftos, personal communication). Matrix GLA protein (MGP) is also a potentially useful biomarker for investigative purposes[233] (Table 35).

| | Correlation coefficient | |
	Simple	Partial
Age (=x) vs. creatinine clearance (=y)	$r_{xy} = -0.67$•	
Age vs. serum BGP (=z)	$r_{xz} = -0.44$•	$r_{xz \cdot y} = -0.32$•
Serum BGP vs. creatinine clearance	$r_{zy} = -0.32$•	$r_{zy \cdot x} = -0.04$•
• $p < 0.001$		

Table 33 Interrelationship of age, renal function and serum BGP[225]

Abundance (mg/g bone)	6.4
Molecular weight	14 000
GLA/molecule	5
Solubility (mg protein/ml)	0.2
Precursor	40 000 Mr (extracellular?)
Extraction from bone	By guanidine after demineralization

Table 35 The principal characteristics of matrix GLA protein (MGP)[227,233]

Status	n	Osteocalcin (ng/ml)	Alkaline phosphatase units · litre^{-1}
Normal adults	90	7.1± 3.0	49 ± 12 (n=150)
Long-term steroid therapy	?	4.3± 0.5	
Hip fracture (pin insertion)	24	19.0±15.7	94 ± 45†
Hip prostheses	13	6.4± 3.1	74 ± 25†
Other fractures	8	14.0±10.6	97 ± 28†
Primary hyperparathyroidism	5	22.0±10.9†	66 ± 49 (n=2)
Secondary hyperparathyroidism	11	51.6±32.9†	149 ± 95 (n=3)
Paget's disease of bone	7	22.7±16.2†	251 ±233
Postmenopausal osteoporosis	?	7.0± 0.7	
Malignant hypercalcaemia	?	2.6± 0.9	
Bone metastases			
No treatment	10	37.5±35.8‡	141 ±131†
Treatment	22	6.1± 6.0	114 ± 69†

Table 34 Serum osteocalcin and alkaline phosphatase measurements in patients with bone-related diseases (based on [226])

All values are the mean ± SD.
Results should be interpreted with extreme caution since they depend on the RIA method employed.
Figures in parentheses indicate the value of n for alkaline phosphatase where this is different from the value for osteocalcin.
† $p < 0.01$ vs. normal adults
‡ $p < 0.05$ vs. normal adults

Other biomarkers

Another useful marker in Paget's disease is procollagen, which can be measured by radioimmunoassay[234].

A number of other substances are also under investigation as potential markers in bone disease. They include osteonectin (which might reflect mineralization activity, see Table 36), bone sialoprotein and bone proteoglycan, all of which normally derive from the organic matrix of bone[235] (Table 37). They also include a number of skeletal regulating factors (cytokines/osteokines, or chondrokines) such as skeletal growth factors[237] and what Baylink (personal communication) calls 'aberrant bone protein'.

Tissue	ng osteonectin per 1000 μg protein
Bone	10−20 000
Skin	43
Tendon	151
Muscle	55
Intestine	82
Spleen	38
Brain	18
Serum*	0.0003

Table 36   Osteonectin level in various tissues (J.D. Termine, personal communication)

* 1 μl serum = 100 μg protein

I. Type-I collagen accounts for 90−95% of the total			
II. Non-collagenous proteins			
	Mr	Composition	Function (in vitro)
● Bone GLA protein (osteocalcin)	5 800+	3 residues of γ-carboxy-glutamic acid (vitamin-K-dependent Ca^{2+}-binding aminoacid)	Unknown
● Osteonectin	32 000*	Single polypeptide, N- and O-linked oligosaccharides, phosphoserine residues	Binds to Ca^{2+}, hydroxyapatite and collagen; promotes mineralization
● Bone-specific sialoprotein	80 000**	50% protein, 50% carbohydrate (14% N-acetylneuroaminic acid, 7% galactosamine, 7% glucosamine)	Unknown
● Bone-specific proteoglycan	80 000−120 000*	1 or 2 chondroitin sulfate chains (40 000); core protein (38 000) with oligo-saccharides	Promotes fibrillogenesis of type-1 collagen
● Other glycoproteins	–	–	–
● Phosphoproteins	–	–	–
● α_2-HS glycoprotein	–	–	–
● Other serum proteins	–	–	–

Table 37   Composition of the organic matrix of bone, showing potential biomarkers for bone metabolism (based on [235] and [236])

* Determined by gel filtration in 4 M guanidine hcl
** Determined from aminoacid sequence data

References

1   Copp DH et al: Evidence for calcitonin – a new hormone from the parathyroid that lowers blood calcium. Endocrinology 1962, 70, 638–49.

2   Kumar MA et al: Further evidence for calcitonin, a rapid-acting hormone which lowers plasma-calcium. Lancet 1963/2, 480–2.

3   Foster GV et al: Thyroid origin of calcitonin. Nature 1964, 202, 1303–5.

4   Foster GV et al: Calcitonin production and the mitochondrion-rich cells of the dog thyroid. Nature 1964, 203, 1029–30.

5   Hirsch PF et al: Thyroid hypocalcemic principle and recurrent laryngeal nerve injury as factors affecting the response to parathyroidectomy in rats. Endocrinology 1963, 73, 244–52.

6   Chambers TJ, Moore A: The sensitivity of isolated osteoclasts to morphological transformation by calcitonin. J Clin Endocrinol Metab 1983, 57, 819–24.

7   Chambers TJ, McSheehy PM, Thomson BM, Fuller K: The effect of calcium-regulating hormones and prostaglandins on bone resorption by osteoclasts disaggregated from neonatal rabbit bones. Endocrinology 1985, 116, 234–9.

8   Chambers TJ, Fuller K, McSheehy PM, Pringle JA: The effects of calcium regulating hormones on bone resorption by isolated human osteoclastoma cells. J Pathol 1985, 145, 297–305.

9   Friedman J, Raisz LG: Thyrocalcitonin: inhibitor of bone resorption in tissue culture. Science 1965, 150, 1465–7.

10 Milhaud G et al: Etude du mécanisme de l'action hypocalcémiante de la thyrocalcitonine. CR Acad Sci Paris 1965, 261, 813–6.

11 Martin TJ, Robinson CJ, MacIntyre I: The mode of action of thyrocalcitonin. Lancet 1966, 1, 900–2.

12 Le Douarin N, Le Lièvre C: Démonstration de l'origine neurale des cellules à calcitonine du corps ultimobranchial chez l'embryon de poulet. CR Acad Sci Paris 1970, 270, 2857–60.

13 Pearse AGE, Polak JM: Cytochemical evidence for the neural crest origin of mammalian ultimobranchial C cells. Histochemie 1971, 27, 96–102.

14 Baber EC: Contributions to the minute anatomy of the thyroid gland. Proc R Soc London 1876, 24, 240–1.

15 Pearse AGE: The thyroid parenchymatous cells of Baber and the nature and function of their C cell successors in thyroid, parathyroid and ultimobranchial bodies. In: Calcitonin, Proc Symp on thyrocalcitonin and the C cells, London 1967. Ed Taylor SW, Heinemann 1968, 98–109.

16 Copp DH et al: Calcitonin – ultimobranchial hormone. In: Calcitonin, Proc Symp on thyrocalcitonin and the C cells, London 1967. Ed Taylor SW, Heinemann 1968, 306–9.

17 Moseley JM, Matthews EW, Breed RH, Galante L, Tse A, MacIntyre I: The ultimobranchial origin of calcitonin. Lancet 1968, 1, 108–10.

18 MacIntyre I, Stevenson JC: Calcitonin, a modern view of its physiological role and interrelation with other hormones. In: Calcitonin 1980, Proc int Symp, Milan 1980. Ed Pecile A, Excerpta Medica 1981, Int Cong Ser 540, 1–10.

19 Cooper CW, Peng TC, Obie JF, Garner SC: Calcitonin-like immunoreactivity in rat and human pituitary glands: histochemical, in vitro, and in vivo studies. Endocrinology 1980, 107, 98–107.

20 Silva OL, Becker KL: Immunoreactive calcitonin in extrathyroid tissues. In: Calcitonin 1980, Proc int Symp, Milan 1980. Ed Pecile A, Excerpta Medica 1981, Int Cong Ser 540, 144–53.

21 Deftos LJ, Burton D, Catherwood BD, Bone HG, Parthemore JG, Guillemin R, Watkins WB, Moore RY: Demonstration by immunoperoxidase histochemistry of calcitonin in the anterior lobe of the rat pituitary. J Clin Endocrinol Metab 1978, 47, 457–60.

22 Deftos LJ, Burton D, Bone HG, Catherwood BD, Parthemore JG, Moore RY, Minick S, Guillemin R: Immunoreactive calcitonin in the intermediate lobe of the pituitary gland. Life Sci 1978, 23, 743–8.

23 Fischer JA, Tobler PH, Kaufmann M, Born W, Henke H, Cooper PE, Sagar SM, Martin JB: Calcitonin: regional distribution of the hormone and its binding sites in the human brain and pituitary. Proc Natl Acad Sci USA 1981, 78, 7801–5.

24 Flynn JJ, Margules DL, Cooper CW: Presence of immunoreactive calcitonin in the hypothalamus and pituitary lobes of rats. Brain Res Bull 1981, 6, 547–9.

25 Milhaud G: Ectopic secretion of calcitonin. In: Calcitonin 1980, Proc int Symp, Milan 1980. Ed Pecile A, Excerpta Medica 1981, Int Cong Ser 540, 154–69.

26 Milhaud G et al: A new chapter in human pathology: calcitonin disorders and therapeutic use. In: Calcium, parathyroid hormone and the calcitonins, Proc 4th parathyroid conf, Chapel Hill (NC) 1971. Ed Talmage RV, Munson PL, Excerpta Medica 1972, 56–70.

27 Silva OL, Becker KL, Primack A, Doppman J, Snider RH: Ectopic secretion of calcitonin by oat-cell carcinoma. N Engl J Med 1974, 290, 1122–4.

28 Milhaud G, Calmette C, Taboulet J, Julienne A, Moukhtar MS: Letter: Hypersecretion of calcitonin in neoplastic conditions. Lancet 1974, 1, 462–3.

29 Coombes RC, Hillyard C, Greenberg PB, MacIntyre I: Plasma immunoreactive calcitonin in patients with nonthyroid tumours. Lancet 1974, 1, 1080–3.

30 Deftos LJ et al: Simultaneous ectopic production of parathyroid hormone (PTH) and calcitonin (CT). Clin Res 1974, 22, 486A.

31 Murad F, Brewer HB Jr, Vaughan M: Effect of thyrocalcitonin on adenosine 3',5'-cyclic phosphate formation by rat kidney and bone. Proc Natl Acad Sci USA 1970, 65, 446–53.

32 Chausmer AB, Stevens MD, Severn C: Autoradiographic evidence for a calcitonin receptor on testicular Leydig cells. Science 1982, 216, 735–6.

33 Findlay DM, Ng KW, Niall M, Martin TJ: Processing of calcitonin and epidermal growth factor after binding to receptors in human breast cancer cells. Biochem J 1982, 206, 343–50.

34 Koida M, Nakamuta H, Furukawa S, Orlowski RC: Abundance and location of 125-I-salmon calcitonin binding site in rat brain. Jpn J Pharmacol 1980, 30, 575–7.

35   Marx SJ et al: Calcitonin receptors of kidney and bone. Science 1972, 178, 999–1001.

36   Maurer R, Marbach P, Mousson R: Salmon calcitonin binding sites in rat pituitary. Brain Res 1983, 261, 346–8.

37   Olgiati VR, Guidobono F, Netti C, Pecile A: Localization of calcitonin binding sites in rat central nervous system: evidence of its neuroactivity. Brain Res 1983, 265, 209–15.

38   Rao LG et al: Immunohistochemical demonstration of calcitonin binding to specific cell types in fixed rat bone tissue. Endocrinology 1981, 108, 1972–78.

39   Chabardes D, Gagnan-Brunette M, Imbert-Teboul M, Gontcharevskaia O, Montegut M, Clique A, Morel F: Adenylate cyclase responsiveness to hormones in various portions of the human nephron. J Clin Invest 1980, 65, 439–48.

40   Chabardes D, Imbert-Teboul M, Montegut M, Clique A, Morel F: Distribution of calcitonin-sensitive adenylate cyclase activity along the rabbit kidney tubule. Proc Natl Acad Sci USA 1976, 73, 3608–12.

41   Henke H, Tobler PH, Fischer JA: Localization of salmon calcitonin binding sites in rat brain by autoradiography. Brain Res 1983, 272, 373–7.

42   Marx SJ, Aurbach GD, Gavin JR, Buell DW: Calcitonin receptors on cultured human lymphocytes. J Biol Chem 1974, 249, 6812–6.

43   Moran J, Hunziker W, Fischer JA: Calcitonin and calcium ionophoresis: cyclic AMP responses in cells of a human lymphoid line. Proc Natl Acad Sci USA 1978, 75, 3984–88.

44   Findlay DM, Michelangeli VP, Eisman JA, Frampton RJ, Moseley JM, MacIntyre I, Whitehead R, Martin TJ: Calcitonin and 1,25–dihydroxyvitamin D3 receptors in human breast cancer cell lines. Cancer Res 1980, 40, 4764–7.

45   Lamp SJ, Findlay DM, Moseley JM, Martin TJ: Calcitonin induction of a persistent activated state of adenylate cyclase in human breast cancer cells. J Biol Chem 1981, 256, 12269–74.

46   Hunt NH, Ellison M, Underwood JC, Martin TJ: Calcitonin-responsive adenylate cyclase in a calcitonin-producing human cancer cell line. Br J Cancer 1977, 35, 777–84.

47   Mundy CR, Altman AJ, Gondek MD, Bandelin JG: Direct resorption of bone by human monocytes. Science 1977, 196, 1109–11.

48   Teitelbaum SL, Stewart CC, Kahn AJ: Rodent peritoneal macrophages as bone resorbing cells. Calcif Tissue Int 1979, 27, 255–61.

49   Chambers TJ, Dunn CJ: Pharmacological control of osteoclastic motility. Calcif Tissue Int 1983, 35, 566–70.

50   Tashjian AH Jr, Wright DR, Ivey JL, Pont A: Calcitonin binding sites in bone: relationships to biological response and "escape". Recent Prog Horm Res 1978, 34, 285–334.

51   Wener JA, Gorton SJ, Raisz LG: Escape from inhibition or resorption in cultures of fetal bone treated with calcitonin and parathyroid hormone. Endocrinology 1972, 90, 752–9.

52   Austin LA, Heath H: Calcitonin: physiology and pathophysiology. N Engl J Med 1981, 304, 269–78.

53   Mazzuoli GF et al: Il fenomeno escape e plateau. In: The effects of calcitonin in man, Proc 1st int Wkshp, Florence 1982. Ed Gennari C, Segre G, Masson 1983, 75–84.

54   Marx SJ, Woodward C, Aurbach GD, Glossmann H, Keutmann HT: Renal receptors for calcitonin. Binding and degradation of hormone. J Biol Chem 1973, 248, 4797–802.

55   Bouizar Z, Rostene WH, Moukhtar MS, Milhaud G: Characterization and quantitative topographical distribution of salmon calcitonin-binding sites in rat kidney sections. FEBS Lett 1986, 196, 19–22.

56   Guttmann S et al: Distribution of calcitonins between their receptors and antibodies. In: The effects of calcitonin in man, Proc 1st int Wkshp, Florence 1982. Ed Gennari C, Segre G, Masson 1983, 25–31.

57   Findlay DM, Michelangeli VP, Moseley JM, Martin TJ: Calcitonin binding and degradation by two cultured human breast cancer cell lines. Biochem J 1981, 196, 513–20.

58   Heynen G, Franchimont P: (The circulating forms of human calcitonin). Pathol Biol (Paris) 1975, 23, 815–9.

59   Roos BA, Parthemore JG, Lee JC, Deftos LJ: Calcitonin heterogeneity: in vivo and in vitro studies. Calcif Tissue Res 1977, 22 Suppl, 298–302.

60   Parthemore JG, Deftos LJ, Bronzert D: The regulation of calcitonin in normal human plasma as assessed by immunoprecipitation and immunoextraction. J Clin Invest 1975, 56, 835–41.

61   Fischer JA, Tobler PH, Henke H, Tschopp FA: Salmon and human calcitonin-like peptides coexist in the human thyroid and brain. J Clin Endocrinol Metab 1983, 57, 1314–6.

62   Heath H et al: Radioimmunoassay of calcitonin in normal human plasma: problems, perspectives and prospects. Biomed Pharmacother 1984, 38, 241–5.

63   Deftos LJ, Weisman MH, Williams GW, Karpf DB, Frumar AM, Davidson BJ, Parthemore JG, Judd HL: Influence of age and sex on plasma calcitonin in human beings. N Engl J Med 1980, 302, 1351–3.

64   Pavlinac DM, Lenhard LW, Parthemore JG, Deftos LJ: Immunoreactive calcitonin in human cerebrospinal fluid. J Clin Endocrinol Metab 1980, 50, 717–20.

65   Hillyard CJ, Cooke TJ, Coombes RC, Evans IM, MacIntyre I: Normal plasma calcitonin: circadian variation and response to stimuli. Clin Endocrinol (Oxf) 1977, 6, 291–8.

66   Body JJ, Heath H: Estimates of circulating monomeric calcitonin: physiological studies in normal and thyroidectomized man. J Clin Endocrinol Metab 1983, 57, 897–903.

67   Kanis JA, Heynen G, Cundy T, Cornet F, Paterson A, Russell RG: An estimate of the endogenous secretion rate of calcitonin in man. Clin Sci 1982, 63, 145–52.

68   Cecchettin M, Tarquini B, Miolo M, Conte N: The endogenous secretion rate of human calcitonin in normal conditions. Biomed Pharmacother 1986, 40, 19–24.

69   Tiegs RD et al: Secretion and metabolism of monomeric human calcitonin: effects of age, sex and thyroid damage. J Bone Min Res 1986, 1, 339–49.

70   Hirsch PF, Munson PL. Thyrocalcitonin. Physiol Rev 1969, 49, 548–622.

71   Snider RH, Silva OL, Moor CF, Becker KL: Immunochemical heterogeneity of calcitonin in man: effect on radioimmunoassay. Clin Chim Acta 1977, 76, 1–14.

72   Heynen G, Franchimont P: Human calcitonin radioimmunoassay in normal and pathological conditions. Eur J Clin Invest 1974, 4, 213–22.

73   Feletti C et al: Calcitonin and uremic osteodystrophy. In: The effects of calcitonin in man, Proc 1st int Wkshp, Florence 1982. Ed Gennari C, Segre G, Masson 1983, 249–57.

74   Copp DH: Parathyroid hormone, calcitonin and calcium homeostasis. In: Parathyroid hormone and thyrocalcitonin (calcitonin), Proc 3rd parathyroid Conf, Montreal 1967. Ed Talmage RV, Bélanger LF. Excerpta Medica 1968, Int Cong Ser 159, 25–39.

75   Potts JT, Jr et al: Control of secretion of parathyroid hormone. In: Parathyroid hormone and thyrocalcitonin (calcitonin), Proc 3rd parathyroid Conf, Montreal 1967. Ed Talmage RV, Bélanger LF. Excerpta Medica 1968, Int Cong Ser 159, 407–16.

76   Care AD et al: The direct measurement of thyrocalcitonin secretion rate *in vivo*. In: Parathyroid hormone and thyrocalcitonin (calcitonin), Proc 3rd parathyroid Conf, Montreal 1967. Ed Talmage RV, Bélanger LF. Excerpta Medica 1968, Int Cong Ser 159, 417–27.

77   Sethi R et al: Effect of meal on serum parathyroid hormone and calcitonin: possible role of secretin. J Clin Endocrinol Metab 1983, 56, 549–52.

78   Care AD et al: Role of pancreozymin-cholecystokinin and structurally related compounds as calcitonin secretogogues. Endocrinology 1971, 89, 262–71.

79   Robinson MF et al: Variation of plasma immunoreactive parathyroid hormone and calcitonin in normal and hyperparathyroid man during daylight hours. J Clin Endocrinol Metab 1982, 55, 538–44.

80   Manolagas SC, Deftos LJ: No diurnal variations in calcitonin and vitamin D2 (letter). N Engl J Med 1985, 312, 122–3.

81   Talmage RV et al: Calcitonin, feeding and calcium conservation. In: Calcitonin 1980, Proc int Symp, Milan 1980. Ed Pecile A, Excerpta Medica 1981, Int Cong Ser 540, 96–109.

82   Palummeri E et al: (Relations between gastrin and calcitonin secretion after a protein meal in young and elderly subjects). G Clin Med 1981, 62, 394–407.

83   Cannarozzi DB, Canale DD, Donabedian RK: Hypercalcitoninemia in infancy. Clin Chim Acta 1976, 66, 387–92.

84   Roos BA, Cooper, CW, Frelinger AL, Deftos LJ: Acute and chronic fluctuations of immunoreactive and biologically active plasma calcitonin in the rat. Endocrinology 1978, 103, 2180–6.

85   Stevenson JC, Evans IM: Pharmacology and therapeutic use of calcitonin. Drugs 1981, 21, 257–72.

86   Heynen G et al: Human calcitonin. Some physiopathological aspects. In: Calcitonin 1980, Proc int Symp, Milan 1980. Ed Pecile A, Excerpta Medica 1981, Int Cong Ser 540, 208–16.

87   Stevenson JC, Abeyasekera G, Hillyard CJ, Phang KG, MacIntyre I, Campbell S, Townsend PT, Young O, Whitehead MI: Calcitonin and the calcium-regulating hormones in postmenopausal women: effect of oestrogens. Lancet 1981, 1, 693–5.

88   Stevenson JC, Hillyard CJ, MacIntyre I, Cooper H, Whitehead MI: A physiological role for calcitonin: protection of the maternal skeleton. Lancet 1979, 2, 769–70.

89   Whitehead M, Lane G, Young O, Campbell S, Abeyasekera G, Hillyard CJ, MacIntyre I, Phang KG, Stevenson JC: Interrelations of calcium-regulating hormones during normal pregnancy. Br Med J (Clin Res) 1981, 283, 10–2.

90   Munson PL et al: Calcitonin: role during lactation. In: Calcitonin 1980, Proc int Symp, Milan 1980. Ed Pecile A, Excerpta Medica 1981, Int Cong Ser 540, 110–22.

91   Toverud SU, Cooper CW, Munson PL: Calcium metabolism during lactation: elevated blood levels of calcitonin. Endocrinology 1978, 103, 472–9.

92   Dupuy B: Antistress effects of calcitonin (editorial). Biomed Pharmacother 1983, 37, 54–7.

93   Vora NM et al: Comparative effect of calcium and of the adrenergic system on calcitonin secretion in man. J Clin Endocrinol Metab 1978, 46, 567–71.

94   Fiore CE et al: Neuroendocrine modulators affecting serum calcitonin levels in man. In: Calcitonin 1980, Proc int Symp, Milan 1980. Ed Pecile A, Excerpta Medica 1981, Int Cong Ser 540, 170–82.

95   Owyang C, Heath H, Sizemore GW, Go VL: Comparison of the effects of pentagastrin and meal-stimulated gastrin on plasma calcitonin in normal man. Am J Dig Dis 1978, 23, 1084–8.

96   Williams GA, Kukreja SC, Sethi R, Hargis GK, Bowser EN: Parathyroid hormone and calcitonin secretion in man: effect of dopamine agonists and antagonists. Horm Metab Res 1986, 18, 64–6.

97   Leggate J, Farish E, Fletcher CD, McIntosh W, Hart DM, Sommerville JM: Calcitonin and postmenopausal osteoporosis. Clin Endocrinol (Oxf) 1984, 20, 85–92.

98   Greenberg C, Kukreja SC, Bowser EN, Hargis GK, Henderson WJ, Williams GA: Effects of estradiol and progesterone on calcitonin secretion. Endocrinology 1986, 118, 2594–8.

99   Christiansen C, Christensen MS, McNair P, Hagen C, Stocklund KE, Transbol I: Prevention of early postmenopausal bone loss: controlled 2–year study in 315 normal females. Eur J Clin Invest 1980, 10, 273–9.

100 Recker RR, Saville PD, Heaney RP: Effect of estrogens and calcium carbonate on bone loss in postmenopausal women. Ann Intern Med 1977, 87, 649–55.

101 Lindsay R, Hart DM, Aitken JM, MacDonald EB, Anderson JB, Clarke AC: Long-term prevention of postmenopausal osteoporosis by oestrogen. Evidence for an increased bone mass after delayed onset of oestrogen treatment. Lancet 1976, 1, 1038–41.

102 Stevenson JC, Whitehead MI: Postmenopausal osteoporosis. Br med J 1982, 285, 585–8.

103 Gordan GS, Picchi J, Roof BS: Antifracture efficacy of long-term estrogens for osteoporosis. Trans Assoc Am Physicians 1973, 86, 326–32.

104   Ziegler R, Raue F: Variations of plasma calcitonin levels measured by radioimmunoassay systems for human calcitonin. Biomed Pharmacother 1984, 38, 245–51.

105   Foresta C, Busnardo B, Ruzza G, Zanatta G, Mioni R: Lower calcitonin levels in young hypogonadic men with osteoporosis. Horm Metab Res 1983, 15, 206–7.

106   Morimoto S, Onishi T, Okada Y, Tanaka K, Tsuji M, Kumahara Y: Comparison of human calcitonin secretion after a 1–minute calcium infusion in young normal and in elderly subjects. Endocrinol Jpn 1979, 26, 207–11.

107   Dupuy B, Angibeau R, Mendoza N, Rouais F, Blanquet P: Lithium effect on calcitonin secretion in the rat. Biomedicine 1980, 33, 207–8.

108   Raue F et al: Acute effect of 1,25–dihydroxyvitamin D3 on calcitonin secretion in rats. Horm metab Res 1983, 15, 208–9.

109   Tagliaro F, Capra F, Dorizzi R, Luisetto G, Accordini A, Renda E, Parolin A: High serum calcitonin levels in heroin addicts. J Endocrinol Invest 1984, 7, 331–3.

110   Stevenson JC, Hillyard CJ: Thyroid cancer: tumour markers. Recent Results Cancer Res 1980, 73, 60–7.

111   Milhaud G, Tubiana M, Parmentier C, Coutris G: (Epithelioma of the thyroid secreting thyrocalcitonin). CR Acad Sci (D) (Paris) 1968, 266, 608–10.

112   Milhaud G: Place de la calcitonine en pathologie et thérapeutique. Actual pharmacol (Paris) 1978, 30, 7–34.

113   Motta L, Maugeri D: La calcitonina nell'anziano. In: The effects of calcitonin in man, Proc 1st int Wkshp, Florence 1982. Ed Gennari C, Segre G, Masson 1983, 267–77.

114   Taggart HM, Chesnut CH, Ivey JL, Baylink DJ, Sisom K, Huber MB, Roos BA: Deficient calcitonin response to calcium stimulation in postmenopausal osteoporosis? Lancet 1982, 1, 475–8.

115   Body JJ, Demeester-Mirkine N, Borkowski A, Suciu S, Corvilain J: Calcitonin deficiency in primary hypothyroidism. J Clin Endocrinol Metab 1986, 62, 700–3.

116   Carey DE et al: Calcitonin secretion in congenital nongoitrous cretinism. J clin Invest 1980, 65, 892–5.

117   Deftos LJ et al: Calcium and skeletal metabolism. West J Med 1975, 123, 447–58.

118   Deftos LJ: Pathophysiology of calcitonin secretion in different species. In: The effects of calcitonin in man, Proc 1st int Wkshp, Florence 1982. Ed Gennari C, Segre G, Masson 1983, 43–53.

119   Weil C: Gastroenteropancreatic endocrine tumours. Klin Wochenschr 1985, 63, 433–59.

120   Pearse AG, Polak JM: Neural crest origin of the endocrine polypeptide (APUD) cells of the gastrointestinal tract and pancreas. Gut 1971, 12, 783–8.

121   Baylin SB: Ectopic production of hormones and other proteins by tumors. Hosp Pract 1975, 10, 117–26.

122   Deftos LJ, McMillan PJ, Sartiano GP, Abuid J, Robinson AG: Simultaneous ectopic production of parathyroid hormone and calcitonin. Metabolism 1976, 25, 543–50.

123   Deftos LJ, Roos BA, Bronzert D, Parthemore JG: Immunochemical heterogeneity of calcitonin in plasma. J Clin Endocrinol Metab 1975, 40, 409–12.

124   Roos BA, Deftos LJ, Roberts G, Wilson P, Bundy L: Radioimmunoassay of calcitonin in plasma, normal thyroid, and medullary thyroid carcinoma of the rat. J Lab Clin Med 1976, 88, 173–82.

125   Hillyard CJ, Coombes RC, Greenberg PB, Galante LS, MacIntyre I: Calcitonin in breast and lung cancer. Clin Endocrinol (Oxf) 1976, 5, 1–8.

126   Wallach SR, Royston I, Taetle R, Wohl H, Deftos LJ: Plasma calcitonin as a marker of disease activity in patients with small cell carcinoma of the lung. J Clin Endocrinol Metab 1981, 53, 602–6.

127   Milhaud G, Calmettes C, Dreyfuss G, Moukhtar MS: An unusual trabecular thyroid cancer producing calcitonin. Experientia 1970, 26, 1381–3.

128   Milhaud G, Calmettes C, Raymond JP, Bignon J, Moukhtar MS: (Thyrocalcitonin-secreting carcinoid). CR Acad Sci (D) (Paris) 1970, 18, 2195–8.

129   Silva OL, Broder LE, Doppman JL, Snider RH, Moore CF, Cohen MH, Becker KL: Calcitonin as a marker for bronchogenic cancer: a prospective study. Cancer 1979, 44, 680–4.

130   Milhaud G et al: Calcitonin. In: Endocrinology, Proc 5th int Cong Endocrinol, Hamburg 1977. Ed James VHT. Excerpta Medica 1977, Int Cong Ser 403, 2, 430–6.

131   Raue F, Bayer JM, Rahn KH, Herfarth C, Minne H, Ziegler R: Hypercalcitoninaemia in patients with pheochromocytoma. Klin Wochenschr 1978, 56, 697–701.

132   Salle BL, David L, Chopard JP, Grafmeyer DC, Renaud H: Prevention of early neonatal hypocalcemia in low birth weight infants with continuous calcium infusion: effect on serum calcium, phosphorus, magnesium, and circulating immunoreactive parathyroid hormone and calcitonin. Pediatr Res 1977, 11, 1180–5.

133   Parthemore JG, Deftos LJ: Calcitonin secretion in normal human subjects. J Clin Endocrinol Metab 1978, 47, 184–8

134   Lambert PW et al: Pre- and postoperative studies of plasma calcitonin in primary hyperparathyroidism. J Clin Invest 1979, 63, 602–8.

135   Morimoto S, Onishi T, Kumahara Y, Avioli L: Differentiation of pseudo- and idiopathic hypoparathyroidism by measuring urinary calcitonin. J Clin Endocrinol Metab 1983, 57, 1216–20.

136   Avioli LV. Editorial: Vitamin D, the kidney and calcium homeostasis. Kidney Int 1972, 2, 241–6.

137   Silva OL et al: Human calcitonin and serum phosphate. Lancet 1974, 1, 1055.

138   Lee JC, Parthemore JG, Deftos LJ: Immunochemical heterogeneity of calcitonin in renal failure. J Clin Endocrinol Metab 1977, 45, 528–33.

139   Heynen G et al: Evidence that endogenous calcitonin protects against renal bone disease. Lancet 1976, 2, 1322–5.

140   Isaac R, Nivez MP, Piamba G, Fillastre JP, Ardaillou R: Influence of calcium infusion on calcitonin and parathyroid hormone concentrations in normal and hemodialyzed subjects. Clin Nephrol 1975, 3, 14–7.

141   Dambacher MA, Hunziker W, Fischer JA: (Significance of plasma calcitonin for the clinical diagnosis). Dtsch Med Wochenschr 1977, 102, 1191–3.

142   Silva OL, Becker KL, Shalhoub RJ, Snider RH, Bivins LE, Moore CF: Calcitonin levels in chronic renal disease. Nephron 1977, 19, 12–8.

143   Nielsen HE, Christensen CK, Olsen KJ: Serum calcitonin in patients with chronic renal disease. Acta Med Scand 1979, 205, 615–8.

144   Korman MG, Laver MC, Hansky J: Hypergastrinaemia in chronic renal failure. Br Med J 1972, 1, 209–10.

145   Ardaillou R, Beaufils M, Nivez MP, Isaac R, Mayaud C, Sraer JD: Increased plasma calcitonin in early acute renal failure. Clin Sci Mol Med 1975, 49, 301–4.

146   Ardaillou R, Sizonenko P, Meyrier A, Vallee G, Beaugas C: Metabolic clearance rate of radioiodinated human calcitonin in man. J Clin Invest 1970, 49, 2345–52.

147   Canale DD, Donabedian RK: Hypercalcitoninemia in acute pancreatitis. J Clin Endocrinol Metab 1975, 40, 738–41.

148   Avioli LV, Birge SJ, Scott S, Shieber W: Role of the thyroid gland during glucagon-induced hypocalcemia in the dog. Am J Physiol 1969, 216, 939–45.

149   Robertson GM Jr et al: Inadequate parathyroid response in acute pancreatitis. New Engl J Med 1976, 294, 512–6.

150   Condon JR, Ives D, Knight MJ, Day J: The aetiology of hypocalcaemia in acute pancreatitis. Br J Surg 1975, 62, 115–8.

151   Norberg HP, DeRoos J, Kaplan EL: Increased parathyroid hormone secretion and hypocalcemia in experimental pancreatitis: necessity for an intact thyroid gland. Surgery 1975, 77, 773–9.

152   Paloyan E et al: The role of glucagon hypersecretion in the relationship of pancreatitis and hyperparathyroidism. Surgery 1967, 62, 167–73.

153   Lawrence AM: Pancreatic alpha-cell function in miscellaneous clinical disorders. In: Glucagon. Ed Lefebvre PJ, Unger RH. Pergamon Press 1972, 259–74.

154   Weir GC, Lesser PB, Drop LJ, Fischer JE, Warshaw AL: The hypocalcemia of acute pancreatitis. Ann Intern Med 1975, 83, 185–9.

155   Sowa M, Appert HE, Howard JM: The hypocalcemic activity of pancreatic tissue homogenate in the dog. Surg Gynecol Obstet 1977, 144, 365–70.

156   Watson CG, Steed DL, Robinson AG, Deftos LJ: The role of calcitonin and parathyroid hormone in the pathogenesis of post-thyroidectomy hypocalcemia. Metabolism 1981, 30, 588–9.

157   Rixon RH, MacManus JP, Whitfield JF: The control of liver regeneration by calcitonin, parathyroid hormone and 1alpha,25–dihydroxycholecalciferol. Mol Cell Endocrinol 1979, 15, 79–89.

158   Dibella FP et al: Serum immunoreactive calcitonin in patients with "toxic shock syndrome". Calcif Tiss Int 1981, 33, 303.

159   de Boer AC, Mulder H, Fischer HR, Schopman W, Hackeng WH, Silberbusch J: Characteristic changes in the concentrations of some peptide hormones, in particular those regulating serum calcium, in acute pancreatitis and myocardial infarction. Acta Med Scand 1981, 209, 193–8.

160   David L et al: Serum immunoreactive calcitonin in low birth weight infants. Description of early changes; effect of intravenous calcium infusion; relationships with early changes in serum calcium, phosphorus, magnesium, parathyroid hormone, and gastrin levels. Pediatr Res 1981, 15, 803–8.

161   Ekeland A et al: Increase in plasma calcitonin following femoral fracture in rats. Acta Orthop Scand 1981, 52, 513–8.

162   Arver S, Bucht E, Sjöberg HE: Calcitonin-like immunoreactivity in human milk, longitudinal alterations and divalent cations. Acta Physiol Scand 1984, 122, 461–4.

163   Baron R, Saffar JL: A quantitative study of the effects of prolonged calcitonin treatment on alveolar bone remodelling in the golden hamster. Calcif Tissue Res 1977, 22, 265–74.

164   Frost HM: Dynamics of bone remodeling. In: Bone dynamics. Ed Frost HM. Little, Brown. 1964, 315–34.

165   Owen M: The origin of bone cells in the postnatal organism. Arthritis Rheum 1980, 23, 1073–80.

166   Owen M: Cellular dynamics of bone. In: The Biochemistry and Physiology of Bone. Ed Bourne GH. Academic Press, 1971, 3, 271–98.

167   Vaughan J: Osteogenesis and haematopoiesis. Lancet 1981, 2, 133–6.

168   Rodan GA, Rodan SB: Expression of the osteoblastic phenotype. In: Bone and mineral research. Ed Peck WA. Elsevier 1984, 244–85.

169   McSheehy PMJ, Chambers TJ: Osteoblast-like cells in the presence of parathyroid hormone release soluble factor that stimulates osteoclastic bone resorption. Endocrinology 1986, 119, 1654–9.

170   Kohler H: Wechselwirkung von Thyreocalcitonin und Parathormon. Schweiz Med Wochenschr 1968, 98, 728.

171   Munson PL: Physiology and pharmacology of thyrocalcitonin. In: Handbook of Physiology. Ed Aurbach GD. American Physiological Society, Washington 1976, 7/7, 443–64.

172   Accardo G et al: Support for the clinical use of calcitonin: electron microscope study of the functional state of bone cells of rats after chronic treatment with calcitonin. Curr Ther Res Clin Exp 1982, 31, 422–3.

173   Wase AW, Solewski J, Rickes E, Seidenberg J: Action of thyrocalcitonin on bone. Nature 1967, 214, 388–9.

174   Parfitt AM: Bone remodeling in the pathogenesis of osteoporosis. Med Times 1981, 109, 80.

175   Borle AB: Regulation of cellular calcium metabolism and calcium transport by calcitonin. J Membr Biol 1975, 21, 125–46.

176   MacIntyre I, Parsons JA: Blood-bone calcium equilibrium in the perfused cat tibia and the effect of thyroid calcitonin. J Physiol Lond 1966, 183, 31–33P.

177   Talmage RV et al: The physiological significance of calcitonin. In: Bone and mineral research 1. Ed Peck WA. Excerpta medica 1983, 74–143.

178   Eilam Y, Szydel N, Harell A: Effects of calcitonin on transport and intracellular distribution of exchangeable Ca^{2+} in primary culture of bone cells. Mol-Cell-Endocrinol 1980, 18, 215–25.

179   Harell A, Binderman I, Guez M: Tissue culture of bone cells: mineral transport, calcification and hormonal effects. Isr J Med Sci 1976, 12, 115–23.

180   Borle AB: Calcium metabolism at the cellular level. Fed Proc 1973, 32, 1944–50.

181   Borle AB. In: Calcium-regulating hormones. Ed Talmage RV, Owen M, Parsons JA. Int Cong Ser 346. Excerpta Medica, 1975, 217–28.

182   Borle AB: Control, modulation, and regulation of cell calcium. Rev Physiol Biochem Pharmacol 1981, 90, 13–15.

183   DiPolo R: Calcium influx in internally dialyzed squid giant axons. J Gen Physiol 1979, 73, 91–113.

184   Denis G, Kuczerpa A: Effect of calcitonin on P uptake in epiphyseal cartilages of P-deficient rats. Can J Physiol Pharmacol 1974, 52, 355–7.

185   Raisz LG: The pharmacology of bone. Introduction. Fed Proc 1970, 29, 1176–8.

186   Ardaillou R: (Renal receptors of parathyroid hormone and calcitonin). Nouv Presse Méd 1978, 7, 4125–30.

187   Heersche JN et al. Calcitonin and the formation of 3', 5'-AMP in bone and kidney. Endocrinology 1974, 94, 241–7.

188   Kawashima H, Torikai S, Kurokawa K: Calcitonin selectively stimulates 25–hydroxyvitamin D3-1alpha-hydroxylase in proximal straight tubule of rat kidney. Nature 1981, 291, 327–9.

189   Caniggia A et al: The rationale of calcitonin treatment in postmenopausal osteoporosis. In: Calcitonin 1980, Proc int Symp, Milan 1980. Ed Pecile A. Excerpta Medica 1981, Int Cong Ser 540, 225–36.

190   Kenny AD: Intestinal calcium absorption and its regulation. CRC Press, Boca Raton, Florida 1981.

191   Harrison HC, Harrison HE: Calcium transport by rat colon in vitro. Am J Physiol 1969, 217, 121–5.

192   Kimberg DV et al: Active transport of calcium by intestine: effects of dietary calcium. Am J Physiol 1961, 200, 1256–62.

193   Bergman L: Studies on early neonatal hypocalcemia. Acta Paediatr Scand (Suppl) 1974 (248), 1–25.

194   Conn RB: A quick reference to normal lab values. Med Times 1979, 107, 77–81.

195   Haynes RC Jr, Murad F: Agents affecting calcification: Calcium, parathyroid hormone, calcitonin, vitamin D, and other compounds. In: The Pharmacological Basis of Therapeutics. Ed Gilman AG et al. MacMillan 1985, 7th ed, 1517–43.

196   Perry HM, Province MA, Droke DM, Kim GS, Shaheb S, Avioli LV: Diurnal variation of serum calcium and phosphorus in postmenopausal women. Calcif Tissue Int 1986, 38, 115–8.

197   Doepfner WEH, Briner U: Calcitonin and gastric secretion. In: Calcitonin 1980, Proc int Symp, Milan 1980. Ed Pecile A. Excerpta Medica 1981, Int Cong Ser 540, 123–35.

198   Pearse AG: The cytochemistry of the thyroid C cells and their relationship to calcitonin. Proc R Soc Lond, 1966, 164, 478–87.

199   Pearse AG, Carvalheira AF: Cytochemical evidence for an ultimobranchial origin of rodent thyroid C cells. Nature 1967, 214, 929–30.

200   Barreras RF: Calcium and gastric secretion. Gastroenterology 1973, 64, 1168–84.

201   Case RM: Calcium and gastrointestinal secretion. Digestion 1973, 8, 269–88.

202   Grodsky GM, Bennett LL: Cation requirements for insulin secretion in the isolated perfused pancreas. Diabetes 1966, 15, 910–3.

203   Nicoletti F, Clementi G, Patti F, Canonico PL, Di Giorgio RM, Matera M, Pennisi G, Angelucci L, Scapagnini U: Effects of calcitonin on rat extrapyramidal motor system: behavioral and biochemical data. Brain Res 1982, 250, 381–5.

204   Rizzo AJ, Goltzman D: Calcitonin receptors in the central nervous system of the rat. Endocrinology 1981, 108, 1672–7.

205   Segre G et al: Non-traditional activities of calcitonin. In: The effects of calcitonin in man, Proc 1st int Wkshp, Florence 1982. Ed Gennari C, Segre G. Masson 1983, 55–64.

206   Deftos LG, Parthemore JG: Effects of age and sex on calcitonin secretion. In: Calcitonin 1980, Proc int Symp, Milan 1980. Ed Pecile A. Excerpta Medica 1981, Int Cong Ser 540, 136–43.

207   Copp DH: Modern view of the physiological role of calcitonin in vertebrates. In: The effects of calcitonin in man, Proc 1st int Wkshp, Florence 1982. Ed Gennari C, Segre G. Masson 1983, 3–12.

208   Salmon DM, Azria M, Zanelli JM: Quantitative cytochemical responses to exogenously administered calcitonins in rat kidney and bone cells. Mol Cell Endocrinol 1983, 33, 293–304.

209   Harell A, Binderman I, Rodan GA: The effect of calcium concentration on calcium uptake by bone cells treated with thyrocalcitonin (TCT) hormone. Endocrinology 1973, 92, 550–5.

210   Deshmukh K et al: Effects of calcitonin and parathyroid hormone on the metabolism of chondrocytes in culture. Biochim Biophys Acta 1977, 499, 28–35.

211   Nijweide PJ, van der Plas A: Regulation of calcium transport in isolated periosteal cells, effects of hormones and metabolic inhibitors. Calcif Tissue Int 1979, 29, 155–61.

212   Borle AB: Effects of thyrocalcitonin on calcium transport in kidney cells. Endocrinology 1969, 85, 194–9.

213   Yamaguchi M, Takei Y, Yamamoto T: Effect of thyrocalcitonin on calcium concentration in liver of intact and thyroparathyroidectomized rats. Endocrinology 1975, 96, 1004–8.

214   Yamaguchi M: Effect of calcitonin on Ca-ATPase activity of plasma membrane in liver of rats. Endocrinol Jpn 1979, 26, 605–9.

215   Yamaguchi M: Regulatory effect of endogenous calcitonin on calcium metabolism in hepatic bile system of rats. Endocrinol Jpn 1980, 27, 381–5.

216 Yamaguchi M et al: Interaction of calcitonin and histamine on bile calcium excretion in thyroparathyroidectomized rats. J Pharmacobiodyn 1981, 4, 418–22.

217 MacManus JP, Whitfield JF: Inhibition by thyrocalcitonin of the mitogenic actions of parathyroid hormone and cyclic adenosine 3',5'-monophosphate on rat thymocytes. Endocrinology 1970, 86, 934–9.

218 Whitfield JF, MacManus JP, Gillan DJ: Inhibition by thyrocalcitonin (calcitonin) of the cyclic AMP-mediated stimulation of thymocyte proliferation by epinephrine. Horm Metab Res 1971, 3, 348–51.

219 Russell RG et al: Biochemical markers of bone turnover in Paget's disease. Metab Bone Dis rel Res 1981, 4/5, 255–62.

220 Chase LR, Aurbach GD: The effect of parathyroid hormone on the concentration of adenosine 3',5'-monophosphate in skeletal tissue in vitro. J Biol Chem 1970, 245, 1520–6.

221 Peck WA: Cyclic AMP as a second messenger in the skeletal actions of parathyroid hormone: a decade-old hypothesis. Calcif Tissue Int 1979, 29, 1–4.

222 Klein DC, Raisz LG: Role of adenosine 3',5'-monophosphate in the hormonal regulation of bone resorption: studies with cultured fetal bone. Endocrinology 1971, 89, 818–26.

223 Varghese Z, Moorhead JF, Wills MR: Plasma hydroxyproline fractions in patients with dialysis osteodystrophy. Clin Chim Acta 1981, 110, 105–11.

224 Krane SM, Munoz AJ, Harris ED Jr: Urinary polypeptides related to collagen synthesis. J Clin Invest 1970, 49, 716–29.

225 Delmas PD, Stenner D, Wahner HW, Mann KG, Riggs BL: Increase in serum bone gamma-carboxyglutamic acid protein with aging in women. Implications for the mechanism of age-related bone loss. J Clin Invest 1983, 71, 1316–21.

226 Slovik DM, Gundberg CM, Neer RM, Lian JB: Clinical evaluation of bone turnover by serum osteocalcin measurements in a hospital setting. J Clin Endocrinol Metab 1984, 59, 228–30.

227 Price PA et al: A new biochemical marker for bone metabolism. Calcif Tissue Int 1979, 28, 159.

228 Rico H et al: Treatment of postmenopausal osteoporosis with calcitonin and calcium. Long-term results. In: Osteoporosis, social and clinical aspects, Proc 2nd int Conf, Athens 1985. Masson 1986, 376–80.

229 Deftos LJ, Parthemore JG, Price PA: Changes in plasma bone GLA protein during treatment of bone disease. Calcif Tissue Int 1982, 34, 121–4.

230 Menon RK et al: Impaired carboxylation of osteocalcin in warfarin-treated patients. J clin Endocrinol Metab 1987, 64, 59–61.

231 Epstein S, Poser J, McClintock R, Johnston CC Jr, Bryce G, Hui S: Differences in serum bone GLA protein with age and sex. Lancet 1984, 1, 307–10.

232 Delmas PD, Wilson DM, Mann KG, Riggs BL: Effect of renal function on plasma levels of bone Gla-protein. J Clin Endocrinol Metab 1983, 57, 1028–30.

233 Price PA: Structure and function of vitamin-K-dependent bone proteins. In: Abstracts volume, Int Symp on Osteoporosis, Aalborg 1987. Ed Jensen J et al, abstract 226.

234 Simon LS, Krane SM, Wortman PD, Krane IM, Kovitz KL: Serum levels of type I and III procollagen fragments in Paget's disease of bone. J Clin Endocrinol Metab 1984, 58, 110–20.

235 Gehron Robey P et al: Bone-specific proteins and their biosynthesis by human bone cells. In: Osteoporosis 1, Proc int Symp, Copenhagen 1984. Ed Christiansen C et al, 441–7.

236 Krane SM: Assessment of mineral and matrix turnover. In: Clinical disorders of bone and mineral metabolism. Ed Frame B, Potts JTjr. Excerpta Medica 1983, 95–8.

237 Farley JR, Baylink DJ: Evidence that skeletal growth factor may be a paracrine effector of bone volume. In: Osteoporosis, Proc int Symp, Copenhagen 1984. Ed Christiansen C et al. Glostrup Hospital Denmark 1984, 423–30.

238 Reginster JY et al: Assessment of the biological effectiveness of nasal synthetic salmon calcitonin (SSCT) by comparison with intramuscular (i.m.) or placebo injection in normal subjects. Bone Mineral 1987, 2, 133–40.

239 Chambers TJ, Azria M: The effect of calcitonin on the osteoclast. Triangle–Sandoz Journal of Medical Science 1988, 27, 53–60.

240 Markowitz M, Rotkin L, Rosen JF: Circadian rhythms of blood minerals in humans. Science 1981, 213, 672–4.

Chapter 3: Exogenous Calcitonin

Foreword
by Professor Leonard J. Deftos

Calcitonin is a unique substance: it is a hormone, it is a tumour marker, it is a drug. It has profound effects on skeletal and mineral metabolism, but it also influences other organ systems. It is one of the three major skeletropic hormones (the others being parathyroid hormone and 1,25-dihydroxycholecalciferol) and it acts in concert with these hormones to regulate mineral metabolism. However, calcitonin also seems to be a member of the neuroendocrine family. It is encoded by a gene which also encodes other neuroendocrine peptides, notably calcitonin gene-related peptide, and it is associated with a substance that is commonly present in neuroendocrine cells, chromogranin A.

The major effect of calcitonin as a drug is to inhibit bone resorption, and it is this action that makes it clinically useful in a variety of bone diseases characterized by increased bone resorption. However, the inhibitory effect of calcitonin on bone resorption is paralleled by its inhibitory effect on a wide variety of organ systems. Two notable examples of such extraskeletal actions of calcitonin are its anti-ulcerogenic effect and its analgesic effect. Although the extraskeletal actions of calcitonin are of considerable clinical importance, it is the skeletal action of the hormone – the inhibition of bone resorption – that dictates its currently established clinical applications. Accordingly, calcitonin is widely used in the treatment of Paget's disease of bone and the hyperresorptive bone disease commonly associated with malignancy. In addition, there is a great deal of interest in the use of calcitonin in the treatment of osteoporosis. This latter indication is attractive because the lower rate of calcitonin secretion in women may contribute to the pathogenesis of osteoporosis. And among the drugs available for the treatment of osteoporosis, calcitonin is very safe.

Although calcitonin is now widely used as a drug, there are three barriers that must be overcome if its clinical usefulness is to be increased – expense, route of administration, and development of resistance. Calcitonin is relatively expensive, but newer methods of synthesis including the use of recombinant DNA should help with this problem. Calcitonin had always to be administered by injection, but new pharmaceutical formulations may obviate this problem; nasal preparations, for instance, are now under clinical study, and even oral forms may soon become feasible with the use of liposomes. New pharmaceutical forms may also help to overcome the problem of resistance to the bioactivity of administered calcitonin. However, a more fundamental approach to this problem is the elucidation of its mechanism of action. It is very likely that resistance to calcitonin involves the interaction of the hormone with its receptor, and understanding the calcitonin-receptor complex should therefore help in overcoming this problem. Molecular biology and gene cloning promise to decipher the mechanism of interaction between calcitonin and its receptor.

Thus, the major barriers to calcitonin's wider application as a drug can be crossed. The next decade of research promises to be even more exciting than the period definitively covered in this book, and calcitonin should enjoy wider use in clinical medicine.

L.J. Deftos

Pharmacological effects and mechanisms of action

The physiological functions of endogenous calcitonin reviewed in the previous chapter can be further elucidated by administering exogenous calcitonin to obtain higher concentrations than those achieved by natural secretion, and this chapter is devoted to these pharmacological aspects of the hormone.

Effects on bone (Fig.50)

As already described in some detail in the preceding chapter, bone is constantly being formed and resorbed in a process known as bone remodelling or turnover. The remodelling process is usually rapid in young people, slower in healthy adults, and either retarded or accelerated in some bone diseases. It may also be affected by a wide variety of drugs and other substances with pharmacological effects (Table 38). One of these is calcitonin, which inhibits resorption[4] of both the inorganic component of bone and the matrix[5-7], principally through an inhibitory effect on the osteoclasts[8]. Calcitonin is also thought to exert a direct or indirect stimulant effect on the osteoblasts, but this action is still poorly understood (Table 39).

Hormones	Other biological agents
Active vitamin-D metabolites [1,25 (OH)$_2$D$_3$]	Albumin
	Bacterial endotoxins
Parathyroid hormone	Prostaglandins
Calcitonin	Osteoclast-activating factor
Corticosteroids	Activated complement factor
Oestrogen	
Androgen	**Drugs**
Insulin	Tetracyclines
Glucagon	Diphosphonates
Thyroxine	Mithramycin
Growth hormone	Thiazides
Gastrin	Salicylates
Secretin	Indomethacin
	Heparin
Endocrine agents and derivatives	Theophylline
	Isoproterenol
Glucocorticoids	Barbiturates
Conjugated oestrogens	Anticonvulsants
Anabolic steroids	Glutethimide
Methyltestosterone	Imidazole
Dihydrotachysterol	Thiophene compounds
5,6-trans-cholecalciferol	Colchicine
5,6-trans-25-hydroxy- cholecalciferol	
1α-hydroxycholecalciferol	**Ions**
	Calcium
	Magnesium
	Phosphate
	Pyrophosphate
	Potassium
	Fluoride

Table 38   Some pharmacological agents with effects on bone[3]

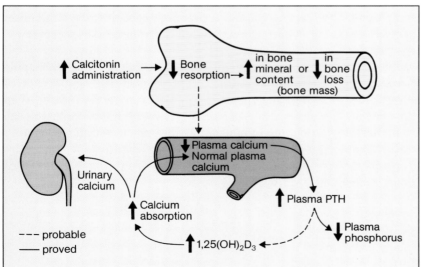

Fig. 50   Possible mechanism of the action of calcitonin on bone (based on [1,2])

Bone resorption	Reversal phase	Bone formation
Inhibition	*Shortening*	*Prolongation?*
Osteoclasts Activity ↓ Lifespan ↓ Numbers ↓ Transfer of Ca and P to mitochondria? Cytoplasmic motility ↓ Brush border ↓		Stimulation Osteoblasts? Chondrogenesis? Mineralization?
Organic matrix Resorption of mineral- ized collagen ↓		
Demineralization ↓		
Effect on local bone factors?		Effect on local bone factors?

Table 39   Known and postulated pharmacological actions of calcitonin in bone

● Bone resorption (anti-osteoclast activity)

Calcitonin counters the osteolytic effects of PTH, vitamin D and other substances, thus slowing accelerated bone remodelling such as occurs, for example, in Paget's disease. It does so by inhibiting the osteoclasts, reducing their activity and motility, their number and the rate at which new ones are formed. These processes and their effects on the organic matrix and the mineralization of bone are supported by a great deal of *in-vitro* and *in-vivo* evidence.

The calvaria of mice aged 5-6 days became manifestly transparent when incubated *in vitro* in a medium containing PTH, an effect which was prevented by calcitonin[9]. The inhibitory effect of calcitonin on bone resorption was also demonstrated in the same model, using mice pretreated with [45]Ca before sacrifice. In the absence of calcitonin the isotope was released into the culture medium during incubation, whereas the presence of calcitonin inhibited its release[10-12] (Fig. 51).

Experiments with bone fragments from rat and mouse embryos[13,14] gave the same results, which have been confirmed by morphological examination[15-19].

Perfusion of isolated cat tibia with a medium containing calcitonin resulted in inhibition of bone resorption, with a net retention of calcium[20].

On the other hand, the inhibitory effect of calcitonin on bone resorption is not always sustained. It was reduced or had disappeared altogether after 12–48 h, for example, in cultures of bone tissue treated with resorption stimulants and in untreated cultures, despite continuous addition of calcitonin in high concentrations[17,21,22]. This 'escape phenomenon'[21] (see next chapter) represents something of an anomaly and must therefore be interpreted with caution. It has been suggested[23] that it might explain why calcitonin

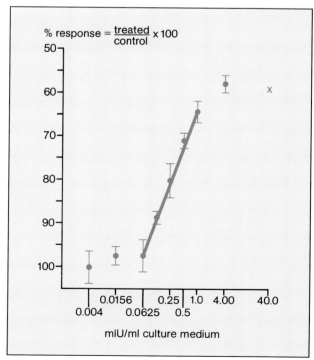

Fig. 51   Dose/response curve for salmon calcitonin in mice. The % response is the ratio of [45]Ca released by treated bones compared with untreated controls over 20 h. Mean values ± SEM[10].

is less effective than was originally hoped in the treatment of hyperparathyroidism, hypercalcaemia of malignancy and other osteolytic disorders, although it does not appear to occur in Paget's disease, where the response of the abnormal osteoclasts to calcitonin is sustained for a long period. This has made calcitonin the mainstay of treatment of this disorder. Also, long-term treatment with calcitonin is associated with an inhibitory effect on hyperactive osteoclasts[24]. This might be explained by the drug's short half-life, which means that with single daily injections the exposure of calcitonin-responsive cells is intermittent, reducing the likelihood of both secondary hyperparathyroidism and down-regulation.

Calcitonin radically altered the internal structure of isolated osteoclasts, inhibiting the cytoplasmic motility (endocytosis, enzymatic exocytosis, cellular interdigitation) that is essential to bone resorption[8].

In populations of bone cells rich in osteoclasts, exposure to calcitonin stimulated receptors linked to adenylate cyclase[25-28] and raised the intracellular concentration of cyclic AMP or its release into the culture medium[22,26] (Table 40). However, it did so by a different mechanism and to a lesser extent than PTH, namely by inhibiting cytoplasmic function in the osteoclasts[18,27,28]. Moreover, this inhibition was not reversed by any of the bone-resorption activators PTH, PGE_2 and $1,25 (OH)_2D_3$.

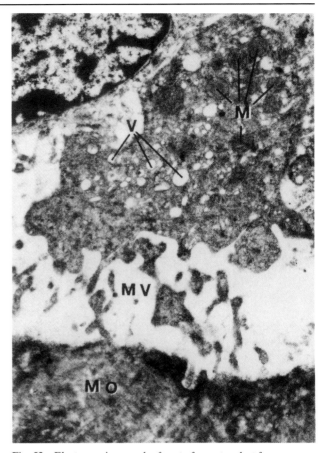

Fig. 52   Electron micrograph of part of an osteoclast from an untreated rat. The "active surface" of the osteoclast adjacent to the bone being resorbed presents elongated microvilli (MV), while its cytoplasm contains numerous vacuoles (V), some of which contain fragments of bone. There are numerous mitochondria (M) but little rough endoplasmic reticulum; MO = bone matrix[29].

Type of CT	Dose (μg/ml)	cAMP (pmol/calvaria)	
		1 Experiment 2	
–	–	6.2 ± 2.3	11 ± 1.6
SCT	0.1	10.0 ± 2.3 †	–
SCT	1.0	14.8 ± 2.3 *	21 ± 1.6 †
HCT	1.0	15.1 ± 2.3 *	19 ± 1.5 †
HCT	5.0	–	20 ± 1.5 ‡

Table 40   Effects of SCT and HCT on cAMP formation by mouse calvaria *in vitro*[22]

Figures are mean values ± SEM
*$p < 0.05$, †$p < 0.01$, ‡$p < 0.001$ vs control

PGI$_2$ and dibutyryl-cyclic AMP had the same effect, probably by raising the intracellular concentration of cyclic AMP[8].

The same effect was also exerted by theophylline, which blocks cyclic AMP degradation. In fact theophylline potentiates the inhibitory effect of calcitonin, dibutyryl-cyclic AMP and PGE_2 on osteoclastic cytoplasmic motility[8].

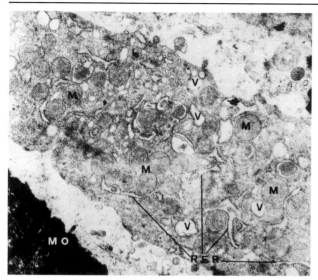

Fig. 53   Part of an osteoclast from an animal treated with calcitonin. The absence of elongated microvilli indicates that the osteoclast is quiescent. Numerous mitochondria, vacuoles and lysosomes can be seen in the cytoplasm, and also more rough endoplasmic reticulum (RER)[29].

It has been postulated that the osteoblasts may produce a factor which stimulates osteoclast activity, reversing the inhibitory effect of calcitonin on the cytoplasm[8].

The inhibitory effect of calcitonin on osteoclast activity has been demonstrated by electron microscopy[29] (Figs 52–53).

Inhibition of bone resorption by calcitonin has also been demonstrated in many in-vivo experiments using histological, morphometric and biochemical (hydroxyproline, bone GLA protein, alkaline phosphatase, cyclic AMP and calcium) criteria[30–33].

Most of these findings have also been confirmed in patients with Paget's disease in studies involving the effect of calcitonin on osteoclast function and numbers and on the organic matrix and the mineralization of bone[34,35].

Effect on the activity and number of osteoclasts

Calcitonin has the following direct and rapid effects on osteoclasts:

In large doses it causes them to start losing their brush border[36] and to move away from the resorption surface[37,38], an effect which occurs in the abnormal osteoclasts of pagetic patients within less than 30 minutes[32].

It shortens their lifespan by an as yet unknown mechanism. It is not clear whether the inhibitory action of calcitonin on the osteoclasts results in a smaller resorption cavity or in slower formation of a cavity of normal size, a distinction which might be crucial to the use of calcitonin in the treatment of osteoporosis.

It reduces the number of osteoclasts, although once again the mechanism involved is still uncertain. Probably it entails decreasing their rate of formation by blocking the fusion of mononuclear marrow cells, the probable committed progenitors of the osteoclasts, which are known to possess calcitonin receptors[15] (Fig. 54).

Effect on the organic matrix of bone

Calcitonin has no effect on the resorption of newly formed and non-mineralized collagen of bone, but it strongly inhibits the resorption of mineralized collagen, probably by preventing the removal of bone salts from mineralized matrix, without directly affecting collagen degradation[5,6].

Effect on bone demineralization

A protective effect of synthetic salmon calcitonin against bone demineralization has been demonstrated in an experimental model of immobilization osteoporosis as well as in paraplegic patients[40]. Further evidence of this effect was provided by another experiment in which plasma levels of radioactivity showed no change in rats given calcitonin less than 24 hours after injection of radioactive calcium. When sufficient time was allowed for the radiocalcium to be incorpo-

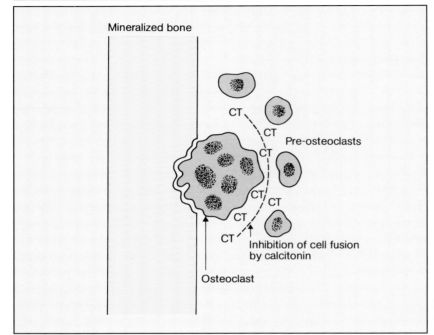

Fig. 54   Osteoclast formation in Paget's disease and the inhibition of cell fusion in the presence of calcitonin[39]

rated into bone, however, calcitonin injection was followed by a rapid fall in both the plasma concentration of radioactive calcium and in the level of total calcium[41].

There is evidence that calcitonin not only inhibits the removal of calcium ions from bone even in the absence of PTH[12,19,37], but also that it actively promotes the transfer of calcium (and phosphate, too, according to some authors[42]) in a labile, easily recoverable form to the skeleton and from the cytosol into the mitochondria. This is supported by findings suggesting that it stimulates the formation of granules in the bone fluid compartment located beneath the surface lining cells and around the osteocytes. These granules might also act as temporary storage for this labile form of calcium phosphate[42], the probable function of which is to prevent both hypercalcaemia and excessive renal excretion of calcium after eating. This would also help to explain the circadian variations in external calcium balance associated with exercise and sleep[43].

● Bone formation

The question of the effect of calcitonin on bone formation is complex and requires more detailed investigation. Indeed, it has not yet been established for certain that it exerts such an effect at all, although there is some experimental evidence for a stimulant effect on the osteoblasts[44], the protein matrix and mineralization – and thus on bone formation.

Studies of the long-term effects of calcitonin on bone remodelling suggest that it lengthens the formation phase at the expense of the reversal phase between resorption and formation[45]. This might be due to an indirect effect via osteoclasts or to a direct effect via osteoblasts. Indeed, since it appears that the two types of cell are very closely interlinked by local factors, it may even act indirectly on either of them via such local factors.

In rats treated with calcitonin, electron microscopy revealed a change in the structure of the osteoblasts, the nucleus becoming larger, the nucleic acid content increasing and the endoplasmic reticulum becoming highly active. These phenomena are indicative of a state in which osteogenesis predominates over osteo-clast activity[29] (Fig. 55).

Repeated administration of calcitonin has been re-ported to increase the growth of cortical bone in young rats and rabbits[46].

Immunocytological evidence has been found for in-ternalization of calcitonin (and PTH) in osteoblasts and direct involvement of calcitonin (and PTH) in os-teoblast regulation, as well as for the absence of

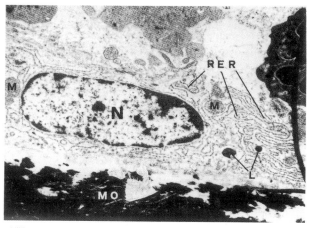

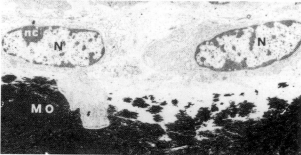

Fig. 55   Osteoblasts near the zone of osteogenesis of the endosteum of rats treated with calcitonin[29]

L = lysosomes, M = mitochondria, N = nucleus,
nc = nucleolus, MO = bone matrix, RER = rough endoplasmic reticulum

both oestradiol and oestradiol receptors in osteo-blasts[47]. However, recent evidence from cell systems *in vitro* suggests that oestrogens may act directly on osteoblasts via specific receptors[48-50].

Calcitonin was found to accelerate repair of experi-mentally induced fractures and to increase heterotopic bone formation in the rat kidney after li-gation of the renal artery[51]. These findings have led to advocacy of the use of calcitonin in the treatment of bone fractures in man, although this application is not generally accepted.

Calcitonin promoted bone-graft survival in a rabbit model in which such grafts were 95% resorbed within 2–3 months under normal circumstances. Moreover, both grafts and donor sites were found on microscopi-cal examination to show more extensive bone forma-tion and better vascularity in treated than in non-treated animals[52].

Calcitonin increased the rate of bone proliferation when applied by means of a collagen sponge to the interior of tooth extraction sites in dogs[53].

It has been reported[54] that the implantation of sponges of gelatin foam impregnated with calcitonin promoted rapid bone repair in patients undergoing surgery for various orthopaedic problems (haemor-rhagic cysts, femur with aseptic necrosis, pseudoar-throsis).

● Extracellular osteogenic substances

The following findings suggest that calcitonin pro-motes the synthesis of extracellular osteogenic sub-stances:

In rabbits with skin lesions, electron microscopy showed that calcitonin increased fibroblast growth and collagen synthesis[55]. By means of electron micro-scopic autoradiography, calcitonin was found to stimulate DNA synthesis in epidermal cells and fi-broblasts during wound healing[56].

Calcitonin promoted the synthesis of glycosaminogly-cans in heterogeneous cultures of femoral epiphyses of bovine foetuses[57].

Calcitonin also increased proteoglycan synthesis in femoral diaphyses from 9-day-old chicken embryos[58].

● Mineralization

Evidence supporting the view that calcitonin actively promotes the mineralization of cartilage and bone, in addition to inhibiting demineralization, includes the following:

Calcitonin-induced hypophosphataemia has been reported to be due both to inhibition of the mobilization of phosphorus from bone and to direct stimulation of phosphorus uptake by bone[59,60].

As has already been said, the plasma level of phosphorus was reported to interact strongly with the hypocalcaemic response to calcitonin[59] and it has been suggested[61] that the primary function of calcitonin might be to regulate phosphorus metabolism, its hypocalcaemic effect in bony vertebrates being only secondary. However, the generally accepted view is that its primary role in mammals is the regulation of extracellular calcium[62].

New light has been shed on the question of whether or not calcitonin's effect on bone formation is independent of its effect on resorption by experiments in rats using implants of demineralized and non-demineralized bone or bone powder which combine the advantages of *in vivo* and *in vitro* methods[63-65]. Demineralized implants induced the formation of new bone at the site of implantation by osteoinduction involving the transformation of host fibroblasts into chondroblasts and then into bone-forming osteoblasts. The process involves genuine ossification, easily distinguishable from dystrophic calcification, and is not followed by bone resorption. It can be quantified by histomorphometry and ^{45}Ca uptake. This type of bone formation is under hormonal control and is promoted by calcitonin[66]. Non-demineralized implants, by contrast, were resorbed and there was no bone formation. This resorption is also under hormonal control and is inhibited by calcitonin. Calcitonin appears to have two effects on induced osteogenesis: it stimulates chondrogenesis and promotes mineralization of the matrix. This interpretation seems to be consistent with other findings in a similar model, showing that calcitonin doubles the rate of [^{3}H]thymidine incorporation by chondroprogenitor cells and increases the rate of synthesis of sulphated proteoglycans by 52%[66].

It has also also been reported that calcitonin strongly promoted mineralization of demineralized bone implants, the greater part of the newly deposited bone salt being found in the implant itself rather than in newly formed bone[63]. This is supported by the finding that calcitonin increased the rate of ^{45}Ca incorporation into matrix-induced tissue by 66%[66], although the technique used in this work was incapable of distinguishing between the mineralization of induced bone and that of the implanted powder.

Summary: Actions of calcitonin at bone level

The pharmacological actions of calcitonin on bone are thought to be:
Inhibition of bone resorption, its primary and undisputed effect, which it achieves by reducing the activity, motility and number of the osteoclasts, possibly by preventing the fusion of mononuclear precursors[67]. This anti-osteolytic effect is then thought to retard bone demineralization and breakdown of the matrix. On the other hand, by depressing osteoclast activity calcitonin may also indirectly depress osteoblast activity. This

could be the reason for the apparent success reported by some authors with an intermittent dosage regimen.

Promotion of bone formation through effects on the osteoblasts, on chondrogenesis and on matrix mineralization[68], although this aspect of its effect requires more extensive investigation, especially in man. It is also possible that it indirectly prolongs the formation phase of the bone remodelling unit's cycle.

Effects on the kidney

Although it is very difficult to obtain a clear picture from the literature of the precise nature of the effect exerted by calcitonin on the kidneys, evidence for the following can be found[69]:

● Urinary excretion (Figs 56–58)

Some authors report that calcitonin increases the urinary excretion of various ions, notably calcium, in man, rat and other animals[71–79], while some report a decrease[80,81] and others very little effect or no effect at all[82–84]. There are also reports for man and rat of a biphasic effect[70,78,85–92] consisting of an initial rise followed by a fall, the calciuric effect of calcitonin being modified by its hypocalcaemic effect and the resultant reduction of the filtered load of calcium. In many cases such discrepancies may be related to the size of the dose administered; for example high doses appear to enhance urinary calcium excretion and low doses to reduce it[80,81]. This calcium-sparing effect is probably due to the promotion of reabsorption in the loop of Henle[80,89,93], particularly in medullary nephrons[93]. The proximal tubule does not seem to be affected, while it is still unclear whether there is any effect on the distal tubule[89,92].

High doses of calcitonin appear to exert a direct and rapid effect on the human kidney, increasing urinary levels of calcium, phosphate, magnesium, sodium, potassium and chloride[74,85,94,95]. This primary effect has been observed in both patients with bone diseases and in healthy subjects[95], and was independent of parathyroid activity[72]. The effects on some of these ions – magnesium, for instance – may be slight and shortlived, but the hypocalcaemic effect does not appear to be destroyed even by nephrectomy.

Calcitonin appears to exert its effect on ion excretion by inhibiting tubular reabsorption or increasing renal clearance[73,75]. Stimulation of cyclic AMP production is probably involved[7,27,28,69,70,72–76,78–81,83–93] and in some species (*not* dog) adenylate-cyclase-linked calcitonin receptors, as distinct from PTH receptors[96], have been identified at various sites in the nephron[27,28,97–99]. High doses of calcitonin raise the urinary excretion of cyclic AMP as well as its renal and plasma concentrations[100,101] (Fig. 59), although its effect on urinary cyclic AMP is only moderate by comparison with that of PTH[75]. This direct action on renal function is independent of its effects on bone, and together these two effects may even give rise to hypocalciuria.

In the rabbit kidney the main sites of calcitonin's activity are the cortical and medullary sections of the thick segment of the ascending limb of the loop of Henle and the proximal part of the distal convoluted tubule. In rats its activity takes place in the distal segment of the nephron, the cortical and medullary parts of the thick ascending loop of Henle, both ends of the distal convoluted tubule, and the cortical segment of the collecting duct[97].

In low doses calcitonin is thought to have little or no effect on the renal excretion of ions[70,80], but there appear to be differences between the different types of calcitonin; salmon calcitonin, for example, is apparently more potent than human or porcine calcitonin in this respect[70]. Also, it should not be forgotten that renal excretion of ions is the result of the combined effects of calcitonin on kidney and on bone, and that other factors are also involved (Table 41).

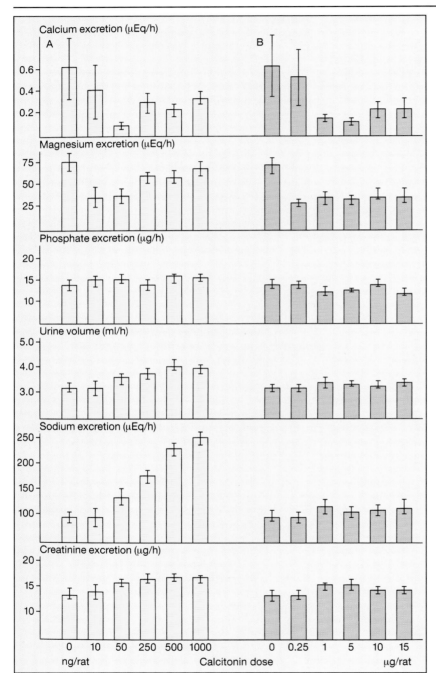

Fig. 56   Effect of (A) salmon and (B) human calcitonin on urine volume, sodium, calcium, magnesium, phosphate and endogenous creatinine excretion[70]

Mean ± SEM, n = 5–8 animals

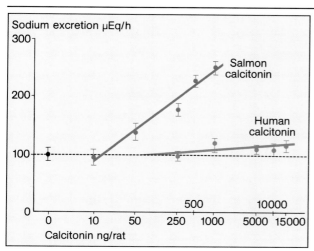

Fig. 57   Regression of sodium excretion on the dose of salmon and human calcitonin[70]

For salmon calcitonin: y = 79.6x + 6.97, r = 0.88, P < 0.001.
For human calcitonin: y = 7.14x + 86.8, r = 0.20, P < 0.2.

Species

+

Type of CT

+

Dose of CT

+

Pathophysiological state of target or other organs involved
(e.g. bone, kidney, gastrointestinal tract)

+

Electrolyte interactions

+

Load of electrolytes (ionic ratios)

+

Effect of CT on tubular reabsorption

+

Duration and sequence of action on kidney, bone and
(possibly) the vascular system

Table 41   Factors which may influence the effect of calcitonin on urinary ion excretion

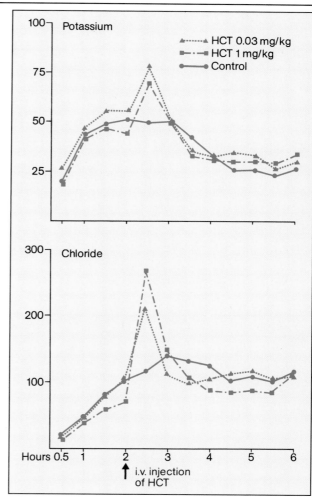

Fig. 58   Time-courses of the urinary excretion of chloride and potassium (μEq per 30 min) after a single intravenous injection of 0.03 and 1 mg/kg human calcitonin (HCT) at 2 h. The rats were infused with Krebs-Ringer bicarbonate buffer solution at a rate of 2 ml/h. Urine samples were collected from the catheterized bladder at 30-minute intervals. Each value represents the mean of 10 rats[71].

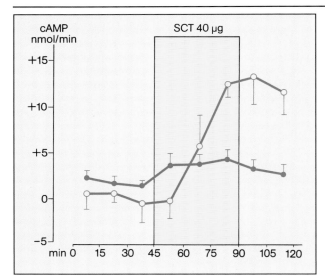

Fig. 59a   Urinary excretion (●) and renal clearance (○) of cAMP in 6 patients[100]

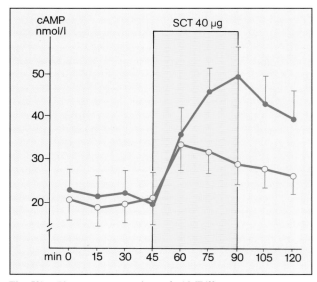

Fig. 59b   Plasma concentrations of cAMP[100]

○ = renal vein, ● = renal artery

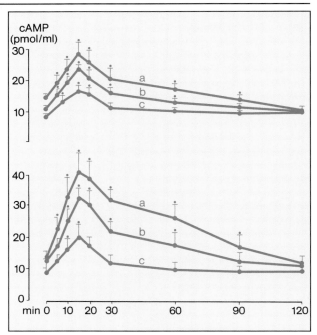

Fig. 59c   Mean (± SEM) cAMP levels in 10 healthy volunteers after 10-minute infusion of SCT(a), HCT(b), and PCT(c)[101]

Upper graph: 50 IU
Lower graph: 100 IU
† indicates a statistically significant difference from baseline

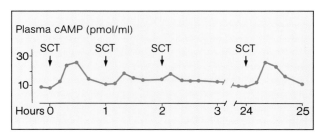

Fig. 59d   Effect of successive infusions of salmon calcitonin on plasma cAMP in a normal subject[101]

Urinary levels of hydroxyproline, which, like serum levels of alkaline phosphatase, are normally raised in conditions of intensive bone-remodelling activity, are lowered by long-term treatment with calcitonin[34] (Fig. 60).

An attempt to systematize these findings, together with further references, will be found in Table 42.

● Renal parenchyma

Calcitonin is thought to reduce the calcium content of the renal parenchyma and to prevent levels of PTH becoming excessive and giving rise to nephrocalcinosis. The mechanism of this action is reported to be the promotion of mitochondrial calcium uptake[113,114], probably via a mechanism involving calcium transport, which is dependent on the local phosphate concentration[113,115,116].

● Vitamin-D metabolism

Calcitonin stimulates the renal production of $1,25(OH)_2D_3$[117], probably by an indirect mechanism involving PTH[118], stimulation of 1α-hydroxylase activity[119,120], and/or an effect on urinary phosphate excretion[121].

Summary: Actions of calcitonin at kidney level

High doses of calcitonin appear to elicit a direct and immediate renal response in the shape of an increase in the urinary excretion of various ions (calcium, phosphate, sodium, potassium, magnesium, chloride), an effect which is then modified as a result of its action on bone. The mechanism probably involves the stimulation of specific adenylate-cyclase-linked receptors triggering an increase in cyclic AMP production. Another effect is thought to be the stimulation of $1,25(OH)_2D_3$ production by the kidney involving

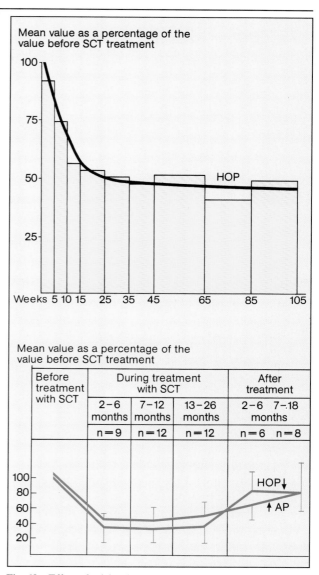

Fig. 60   Effect of calcitonin on urinary hydroxyproline (HOP) and serum alkaline phosphatase (AP)

PTH, 1α-hydroxylase, and/or other mechanisms, although the clinical relevance of these renal effects is only marginal[117]. In pagetic patients calcitonin indirectly reduces raised levels of urinary hydroxyproline.

Substance	Direction of change	Comments		Other References
Calcium	↑(1)↓(2)	1=Immediate renal effect 2=Effect in bone	Single dose: immediate rise[95], then fall to below baseline level[86] Multiple doses: apparent decrease[85], especially in pagetic patients[103,104] and immobilized children[105]	106 107
Phosphate	↑	Reduced reabsorption in the proximal tubule of the nephron; this is also considered to be the site of cellular alkaline phosphatase activity[108]		70,72,75, 77,78,82, 87,88,90, 106,109, 110
Magnesium	↑ ↓	Initial increase, then decrease. Hypomagnesuria in the rat as a result of reabsorption in the thick ascending limb[83], and in pagetic patients?[95]		
Sodium	↑			70,94,111
Potassium	↑	Changes in K^+ level parallel changes in phosphate concentration		95,112
Chloride	↑			94
Water	↑	Diuretic effect is generally slight, but somewhat greater with SCT than HCT		70
Urea	↑?			70,72,75
Hydroxyproline	↓	A reflection of the activity in bone; mainly in pathological states after long-term treatment (in pagetic patients)		34,35
cAMP	↑	Slightly raised urinary excretion; stimulation of CT-sensitive adenylate cyclase		99,100

Table 42   The effect of calcitonin in pharmacological doses on urinary volume and excretion levels of various substances

Effects on the gastrointestinal tract and accessory organs

Besides the physiological evidence of interactions between calcitonin and gastrointestinal function (see Chapter 2), pharmacological studies show that the principal effect of calcitonin at normal doses is a general inhibition of gastric and pancreatic secretory activity. Moreover, the hormone-secreting cells of the gut and its accessory organs belong to the same APUD cell system as the calcitonin-secreting C cells of the thyroid and are closely related to them phylogenetically[122].

● The stomach (Fig.61)

As pharmacological work has clearly shown, calcitonin inhibits gastric-acid secretion, with both preventive and healing effects on experimentally induced ulcers[123–134] (Table 43).

In animals

The pH of the gastric mucosa in mice rose after administration of calcitonin. This effect was potent and dose-dependent above a threshold of 4 IU/kg[123] (Fig. 62).

In conscious rats and cats pentagastrin-induced gastric-acid secretion and ulcer formation were inhibited by concomitant administration of salmon calcitonin (Figs 63 and 64). The inhibition of secretion was dose-dependent over the range 0.04-4 IU/kg/h and correlated with the inhibitory effect on ulceration[123–125].

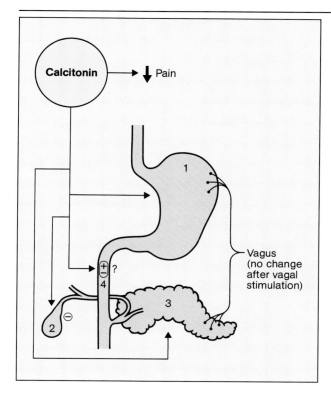

1 *Stomach*
- Inhibition of both *normal* and *pathological* gastric juice secretion (decrease in volume and in H^+, Cl^- and pepsin content), and of pentagastrin-, calcium- and test-meal-stimulated secretion
- Inhibition of gastrin secretion
- Prevention or treatment of experimentally induced ulcers (Shay method; restraint, phenylbutazone, cysteamine)

2 *Gall bladder*
- Inhibition of cholecystokinin-induced contractions

3 *Pancreas*
- Inhibition of normal and pathological secretion and of secretin- and pancreozymin-induced secretion (decrease in amylase concentration)

4 *Intestine*
- Little or no direct effect on calcium absorption at normal calcitonin dose levels
- At high doses, calcitonin causes water and sodium, potassium and chloride ions to be secreted into the intestinal lumen
- Where there is an effect, it may be mediated by stimulation of 1α-hydroxylase secretion, which raises concentrations of the active metabolites of vitamin D.

Fig. 61   Principal pharmacological effects of calcitonin on the gastrointestinal tract and accessory organs (in man and animals)

Effect	Species	Model
Basal secretion		
Inhibition of volume secreted	Man, rat, mouse	
Increase in pH	Mouse	
Stimulated secretion		
Inhibition of volume secreted	Man, rat, cat	Pentagastrin induction, pylorus ligation
Inhibition of free and total acid	Rat, cat	Pentagastrin induction, pylorus ligation
Inhibition of chloride	Cat	Pentagastrin induction
Inhibition of pepsin	Rat	Pylorus ligation
Stimulated ulceration		
Prevention	Rat	Pylorus ligation
		Cysteamine induction
		Phenylbutazone induction
		Stress induction
	Cat	Pentagastrin induction
Treatment/cure	Rat	Phenylbutazone induction

Table 43   Effects of calcitonin on gastric-acid production and/or ulcer formation[123–134]

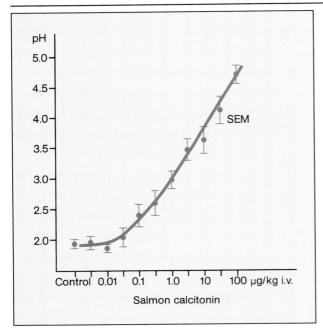

Fig. 62   Effect of synthetic salmon calcitonin on the pH of the gastric mucosa in the mouse[123]

Similar results were obtained in the pylorus-ligated rat model[123–125] (Fig. 65).

Other experimentally induced ulcers (by stress due to restraint, cysteamine or phenylbutazone) were almost completely prevented (inhibitory effect) or healed (curative effect) by calcitonin[123,125] (Figs 66–68).

In man
Calcitonin (single intravenous infusion or orally) has been shown to cause dose-dependent inhibition of gastric-acid secretion (its volume and its free-acid, total-acid, chloride and pepsin content)[126–129]. Both normal secretion and secretion artificially induced by pentagastrin, calcium salts or protein-rich food were inhibited and the effect was seen in healthy volunteers as well as in patients with peptic ulcer[126,127,130–132], Zollinger-Ellison syndrome or gastric hyperacidity[133],

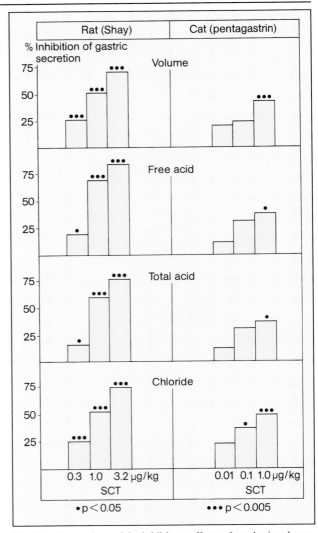

Fig. 63   Comparison of the inhibitory effects of synthetic salmon calcitonin on the volume and the free-acid, total-acid and chloride content of gastric secretions in rats and cats[123]

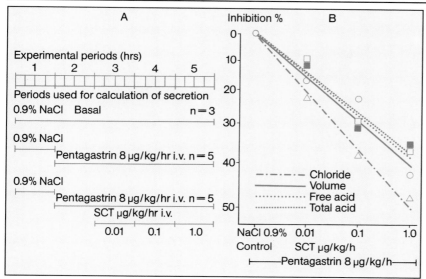

Fig. 64   Inhibitory effect of synthetic salmon calcitonin (SCT) on pentagastrin-stimulated gastric acid secretion in conscious cats with a gastric fistula and catheterized jugular vein. (A) Procedure, (B) Inhibition induced by SCT, as indicated by decreases in volume and in chloride, free-acid and total-acid levels (mean values and regression lines)[123]

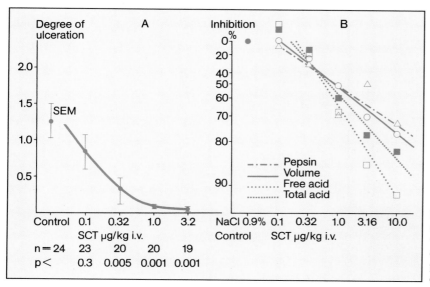

Fig. 65   Inhibition by synthetic salmon calcitonin of ulcer formation (A) and of the volume and the pepsin, free-acid and total-acid content of gastric-acid secretion (B) in pylorus-ligated (Shay) rats. Duration of the experiment: 6 hours[123]

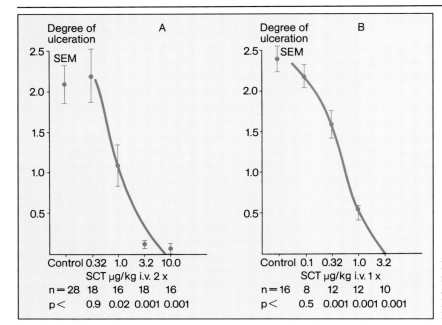

Fig. 66   Inhibition of ulcer formation in rats induced by (A) cysteamine 425 mg/kg s.c. and (B) phenylbutazone 150 mg/kg orally. Duration of the experiments: A = 24 h and B = 6 h[123]

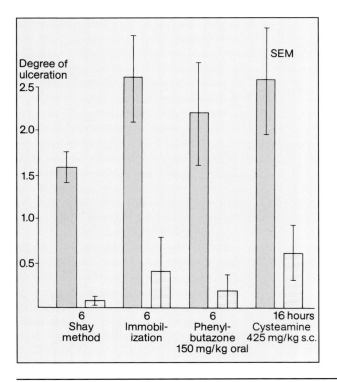

and primary hyperparathyroidism[132,134]. High gastrin levels are also reported to be lowered[123,128]. In man, the inhibitory effect of oral calcitonin on gastric secretion was not accompanied by a hypocalcaemic effect[126], a finding personally confirmed by the author in rats (unpublished).

● The gallbladder

High doses of calcitonin have been reported to inhibit fat-meal or cholecystokinin-induced contraction of the gallbladder[129,135,136], although this is not entirely consistent with the finding that normal doses do not appear to affect gastrointestinal motility[136].

Fig. 67   Inhibitory effect of synthetic salmon calcitonin (3.2 μg/kg i.v.) on ulcer formation in 4 different rat models. The blue columns represent untreated controls and the yellow columns the treated animals[123].

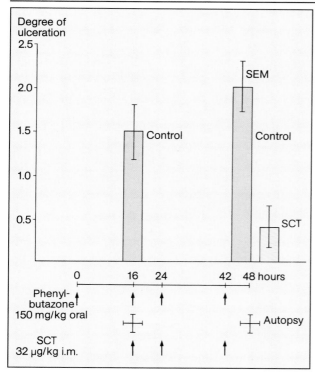

Fig. 68   Healing effect of synthetic salmon calcitonin (SCT) on ulcers induced and maintained by oral phenylbutazone 150 mg/kg. SCT was started 16 h after the onset of ulceration. The blue columns represent untreated controls and the yellow column animals receiving SCT[123].

● The pancreas

Exocrine secretion

Exogenous calcitonin inhibits pancreatic enzyme production, whether basal, artificially induced – by secretin, cholecystokinin/pancreozymin, caerulein or calcium infusion[137,138] – or pathological, as in disorders such as acute pancreatitis in which there is a fall in serum amylase[139,140]. Intravenous infusion at doses between 0.12 and 0.2 IU/kg/h (SCT) inhibited secretory activity by as much as 50%.

However, calcitonin did not affect the level of exocrine secretory activity of the pancreas after dopa-

minergic or cholinergic stimulation[131] or stimulation via the vagus, by insulin or carbamyl choline[140]. Nor does it appear to affect the volume of secretion or the quantity of bicarbonate secreted[141,142]. In the cat the effects of exogenous calcitonin on pancreatic secretion are the same as in man, except that there is a substantial decrease in both the volume secreted and in its bicarbonate content in response to the infusion of calcium[143].

Endocrine secretion

The effect of exogenous calcitonin on the endocrine pancreas is a subject of ongoing controversy. On the one hand calcitonin has been claimed to interfere with glucagon secretion and to some extent with both the release and the effect of somatostatin; on the other hand a slowing effect on the assimilation of glucose and an inhibitory effect on insulin secretion have been reported both in rats given high doses and in the isolated rat pancreas[144,145]. In some pagetic patients, too, blood glucose may rise initially in response to calcitonin (100 IU SCT given s.c.), and glucose tolerance is also affected in some normal subjects.

This slowing of glucose metabolism[146,147] appears to be caused by a temporary reduction in insulin secretion resulting from the hypocalcaemic response to calcitonin[148,149]. This is not a diabetogenic effect but simply an adaptive response, as shown by the fact that it does not persist in pagetic patients on long-term calcitonin. Moreover, the incidence of diabetic symptoms does not appear to be higher in patients with medullary carcinoma of the thyroid than in a normal control population[150].

● The gut

In man, rabbits and rats very high doses of calcitonin increase the secretion of water and of sodium, potassium and chloride ions into the intestinal lumen[151–153]. This effect, which is negligible at normal therapeutic doses[152–159], may explain phenomena such as the diarrhoea which affects patients with medullary carcinoma of the thyroid but which disappears after tumour resection[160]. Infusion of plasma from such patients into dogs has been reported to increase secretion of water into the jejunum[161].

The role of calcitonin in the intestinal absorption of calcium is complex, and reported findings are inconsistent. This is because the effects of calcitonin are influenced by many factors such as dose, species, the existing blood level of calcium, the quantity of calcium supplied by the diet, and the production of active vitamin-D metabolites. However, the following facts have been established:
- Very high doses of calcitonin cause an initial rise in intestinal absorption of calcium[162,163].
- Low doses and physiological levels have no significant effect on absorption except, possibly, to slow it very slightly[152,154–159,164], especially with long-term administration[165,166].

These findings concern calcium absorption from the ileum, jejunum and duodenum and have been obtained from studies in man, dogs, chickens, rabbits, sheep and newborn and adult pigs[151,152,155,157, 164,166,167], but principally in rats[153,154,156,158,159,162,163].

Where it does increase the intestinal absorption of calcium ions, calcitonin seems to act by increasing the formation of active vitamin-D metabolites, which promote calcium uptake. It probably does this by indirectly increasing production of 1,25-dihydroxycholecalciferol from its substrate 25-hydroxycholecalciferol. In vitamin-D-deficient rats, relatively low intravenous doses of calcitonin did not increase the quantity of calcium absorbed[168].

Thus, at normal doses, calcitonin probably has little or no direct effect of any consequence on the intestinal absorption of calcium[7,152,153–159,164,166,167,169]. If such an effect were to be found, it would almost certainly involve 1α-hydroxylase stimulation and increased production of active vitamin-D metabolites.

● Possible mechanisms of action of calcitonin on the gastrointestinal tract

Several hypotheses have been proposed to explain the gastrointestinal role, or roles, of calcitonin, but there is no consensus. One theory, based on the observation that the gastric and pancreatic effects of calcitonin are not antagonized by intravenous calcium, suggests that a direct interaction with gastrin and the pancreozymin/cholecystokinin system at target-cell receptor level is involved[170]. Intragastric and oral administration of calcitonin in rats are also effective, suggesting a direct effect.

Another hypothesis suggests that the effect of calcitonin on the stomach is mediated by somatostatin, release of which it stimulates in a dose-dependent manner[140]. On the other hand calcitonin might act via a central mechanism since, in rats, intraventricular administration inhibited gastric secretion at doses 1000 times lower than the intravenous doses required to produce the same effect. This dose ratio is consistent with findings for other peptides, namely that only a small proportion of an intravenous dose crosses the blood-brain barrier[171].

The potential effects of calcitonin on the turnover and the content of biogenic amines in the stomach[123] have received little attention. It seems, however, that serotonin has the same gastric effects as calcitonin (secretory diarrhoea[172] and inhibition of gastric secretion[173]), which suggests that it may act by liberating serotonin. This appears to be supported by a report that a single calcitonin injection reduced serotonin concentrations in the antrum, duodenum and ileum of rats[174], although other evidence (author's unpublished findings) indicates that serotonin is not involved.

Calcitonin does not affect acetylcholinesterase activity in the stomach[175] and no detailed study has been made of its interactions with the cholinergic system, which plays such an important part in gastric secretion. Nor do we really know what effect it might have on the catecholamine system of the stomach[123].

Recently it has been suggested that the amino-suberic[1,7] analogue of eel calcitonin has antisecretory and (possibly) anti-ulcer activity which is centrally mediated via prostaglandin modulation[176].

● Possible therapeutic implications

Symptomatic improvement (pain relief) was reportedly achieved with calcitonin in exploratory clinical studies to assess its effect on gastric and pancreatic secretion in such disorders as peptic ulcer and gastrointestinal bleeding[133]. In acute pancreatitis before surgical intervention, promising results (pain relief, normalization of enzyme secretion) have also been obtained[139,170,177–179], albeit without a reduction in mortality. Thus, further work is obviously required before any definitive statement can be made.

Some side effects of calcitonin (especially diarrhoea) might also be explained by its actions on the gastrointestinal tract[180].

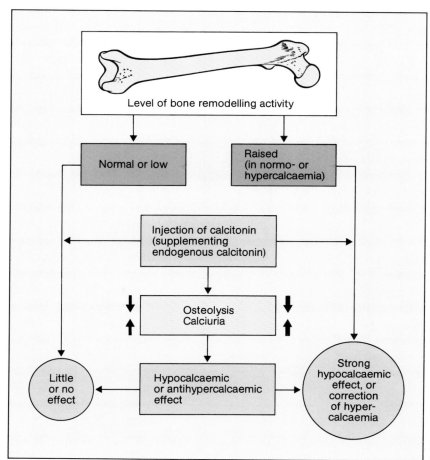

Fig. 69   Mechanisms involved in the effect of calcitonin on blood calcium level.

Summary: Actions of calcitonin at GI level

The primary action of calcitonin on the gastrointestinal tract and its accessory organs appears to consist in inhibition of the secretory activity of the stomach and pancreas, suggesting that it might have potential in the prevention, and even treatment, of peptic ulcer. It also increases the secretion of water and of sodium, potassium, chloride and other ions in the gut, which might explain side effects like diarrhoea which are associated with high doses. At the doses currently used in bone diseases, calcitonin does not appear to affect gastrointestinal calcium absorption, and generally it seems unlikely that it plays a significant gastrointestinal role. At slightly higher doses it is already finding clinical application in acute pancreatitis, in which it relieves pain and restores enzyme production to normal levels.

Effects on blood chemistry

The principal effects of a parenteral dose of calcitonin on blood chemistry in both animals and man are hypocalcaemia (Fig. 69) and hypophosphataemia. These effects are due mainly to calcitonin's anti-osteolytic action, but also to its action on the kidney and, to a much smaller extent, on the gastrointestinal tract (see above).

The magnitude and duration of the *hypocalcaemic effect* in the rat and mini-pig depend on the dose[71,181] (Figs 70 and 71), on the concentrations of other ions present (especially phosphate) and on interactions with other hormones (including PTH). In man this effect may last for several hours, reaching a peak between 3 and 7 hours after administration ([101,182] and A. Fournié, personal communication 1982). The effect also depends on the level of bone remodelling activity[183]. If this is low or normal, as in healthy adults, the hypocalcaemic response is slight, blood calcium falling by only 3–5 mg/l by 1 hour after administration. If, on the other hand, the level of remodelling activity

is high, as in young animals (Table 44), in children, and in patients with bone diseases such as Paget's disease, the hypocalcaemic – or 'antihypercalcaemic' – effect is much more marked (Fig. 69). In children and in patients with Paget's disease (Figs 72–74) the fall in blood calcium level may be by as much as 15 mg/l [186,187], and normal levels may be attained even in infants with idiopathic hypercalcaemia due to vitamin-D hypersensitivity[188]. Similar results have also been obtained in hypercalcaemia following vitamin-D overdosage[186]. Hypercalcaemia due to hyperparathyroidism or bone metastases also responds to

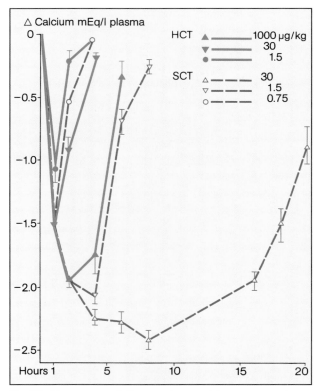

Fig. 70   Time course of effect on plasma calcium concentration (mEq/l) of single intravenous injections of human (HCT) and salmon (SCT) calcitonins, both at three dose levels. Each point represents the mean for 5 rats ± SEM[71].

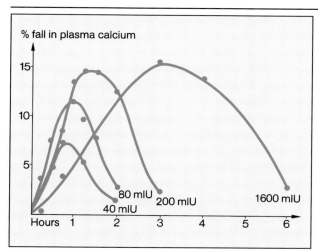

Fig. 71a   Kinetic curves for plasma calcium after injection of calcitonin at various dose levels (mIU · kg⁻¹)

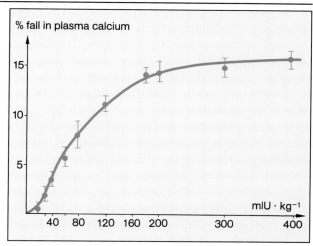

Fig. 71b   Dose-effect curve at 1 h 30 min after calcitonin injection

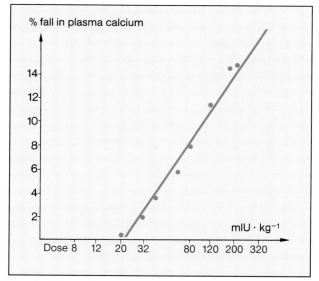

Fig. 71c   Logarithmic dose-effect curve at 1 h 30 min after calcitonin injection

Age of the animals	Fall % (mean ± SEM for 5 animals)
45 days	12.25 ± 1.8
3 months	8.25 ± 0.6
4 months	8.00 ± 0.6
5 months	8.11 ± 0.5
7 months	8.55 ± 0.5
18 months	5.35 ± 0.9

Table 44   Percentage fall in plasma calcium in minipigs given 80 IU/kg salmon calcitonin i.v.[181]

Fig. 71   Effect of i.v. SCT on blood calcium levels in minipigs[181]

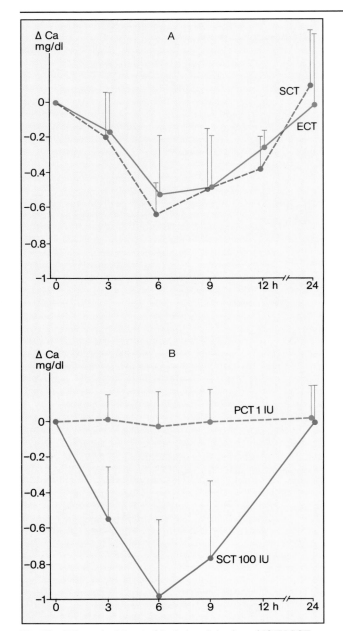

Fig. 72   Effect of calcitonin (single i.m. injection of 80 IU SCT or ECT[A] or of 100 IU SCT or 1 IU PCT[B]) on serum calcium in patients with Paget's disease (n = 7–10) [184,185]

calcitonin, but to a lesser extent (Fig. 75). Reductions of 10–15 mg/l in blood calcium level have been reported, but normal values are very rarely restored [186,189]. On the other hand, the effect of calcitonin is greater the higher the calcium ion concentration before treatment [183,189], although in the author's experience, ionized calcium responds more readily than total calcium. Also, continuous intravenous infusion of calcium over 2–3 days may lead to an attenuation of the hypocalcaemic effect.

The *hypophosphataemic* effect of calcitonin (Fig. 76), which is the result of the lower rate of bone resorption combined with the hyperphosphaturic effect, increases with the dose [71,186].

A number of blood chemistry variables which act as markers for particular diseases exhibit changes after calcitonin administration. The *cyclic AMP* level, for example, was raised [101], whereas the raised *alkaline phosphatase* level characteristic of pagetic patients was lowered or restored to normal by administration of

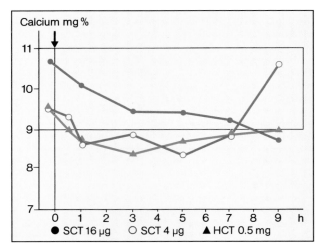

Fig. 73   Reduction in plasma calcium levels after 0.5 mg human calcitonin, compared with effect after 4 or 16 μg synthetic salmon calcitonin in immobilized children (n = 9) with incipient osteoporosis (based on [105])

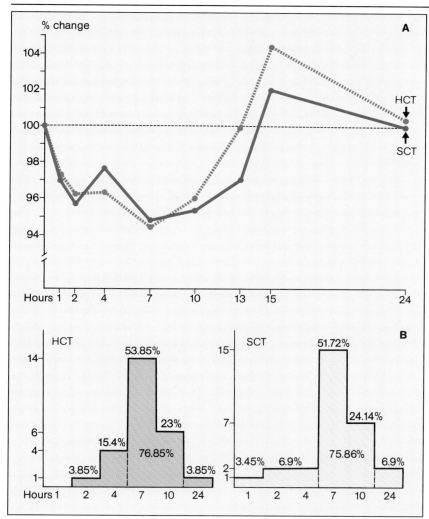

Fig. 74 Comparison of the hypocalcaemic effects in pagetic patients of i.m. administration of 80 IU salmon calcitonin and 0.5 mg human calcitonin (n = 24) [182]

A = Acute hypocalcaemic response test
B = Histograms of time to peak effect

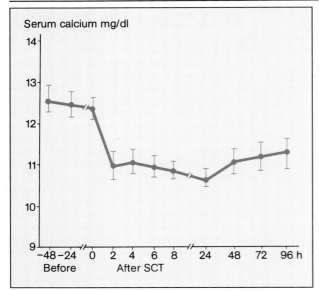

Fig. 75   Mean serum calcium levels (± SEM) for 24 hypercal-caemic patients treated with salmon calcitonin. The 0, 24, 48, 72, and 96 h samples were taken before the patient recieved the morning dose of SCT (4–8 IU/kg every 6–12 h)[189].

calcitonin over a period ranging from a few weeks to 3–6 months[186]. *Plasma urea* levels (Fig. 77) fell in rats given various dosages of salmon or human calcito-nin[71], as did the raised blood levels of *osteocalcine* (bone GLA protein or BGP) occurring in bone dis-eases[190], although the significance of these some-times contradictory findings is not yet precisely under-stood. The raised level of *serum amylase* encountered in acute pancreatitis was also lowered by calcitonin[139].

These findings are summarized and further references given in Table 45.

Effects on the central nervous system

Endogenous immunoreactive calcitonin is found in the central nervous system, where it must therefore be assumed to have some 'non-bone' function. It has also been detected at CNS sites following intramus-cular, intravenous and subcutaneous administration, but only in very low concentration: for example, one-fortieth of an intravenous dose of 100 IU salmon cal-citonin entered the cerebrospinal fluid of anaes-

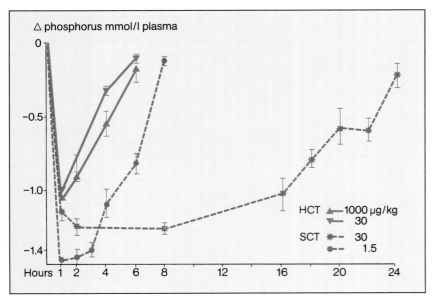

HCT ▲—1000 µg/kg
▼— 30
SCT ✳— 30
●— 1.5

Fig. 76   Time course of effect on plasma phosphorus concentration (mmol/l) of single intravenous injections of human (HCT) and salmon (SCT) calcitonins, both at two dose levels. Each point represents the mean for 5 rats ± SEM[71].

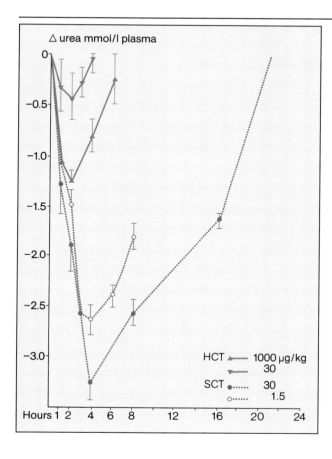

Fig. 77 Time course of plasma urea concentration (mmol/l) after a single intravenous injection of human (HCT) and salmon (SCT) calcitonin, both at 2 dose levels. Each point represents the mean for 5 rats ± SEM[71]

thetized rabbits (M. Lemaire et al., personal communication; Fig. 78). Direct CNS administration has also been used, to man by peridural/epidural, intrathecal, intracisternal and intraventricular routes, and to animals by the intracerebroventricular route. The effects of the latter included reduced prolactin and milk secretion[193], inhibition of gastric acid secretion[171], and appetite suppression[194–196], as well as effects on the actions and levels of some neurotransmitters and their metabolites[197]. In man, calcitonin reportedly alleviates pain associated with bone disorders[198] – but also non-osteogenic pain such as after hysterectomy[199] – and analgesia is still generally regarded as its principal central effect[199–211].

Parameter	Effect	Comments	References
Calcium	↓	Total, ionized (more sensitive and relevant)	71, 101, 105, 181–189
Phosphate	↓		71, 111, 183, 186
Magnesium	Ø ↓	Hypomagnesaemia in man? No change in rats	183
Bone GLA protein	↓	Reduction of the high levels associated with bone diseases	191
Cyclic AMP	↑	At high doses, in part via PTH?	101
Skeletal alkaline phosphatase	↓	After long-term treatment	186, 192
Serum amylase	↓	In acute pancreatitis	139
1,25 (OH)$_2$D$_3$	↑	Through stimulation of renal 1α-hydroxylase	119
Urea	↓	In rats	71

Table 45 Effects of exogenous calcitonin on blood chemistry parameters (man and animals)

↓ = causes a fall   ↑ = causes a rise
Ø = has no effect

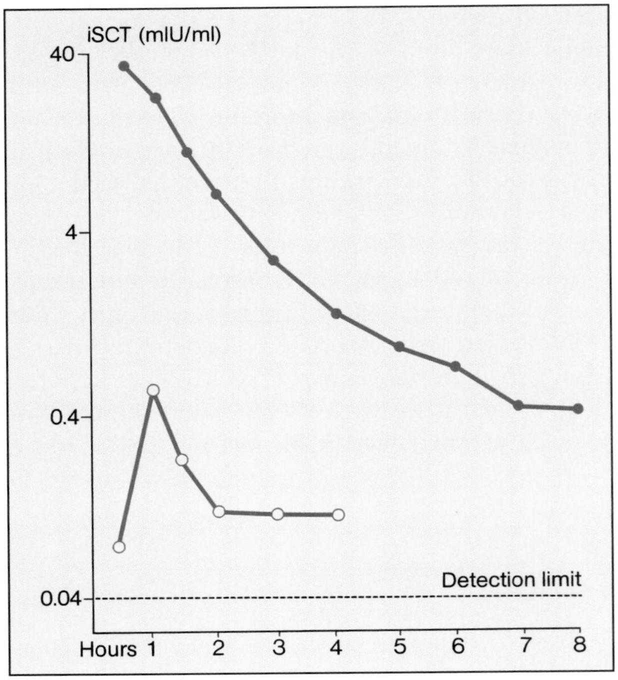

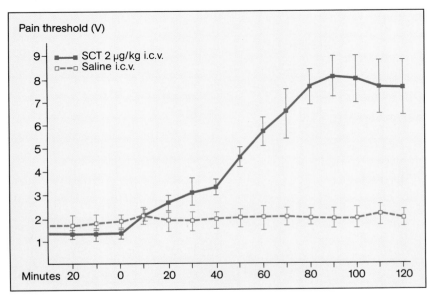

Fig. 78   Concentrations of calcitonin in plasma (●) and CSF (○) of rabbits (n = 5) after an i.v. dose of 100 IU/animal

● Analgesic effect

Although some of the evidence for the analgesic effect of calcitonin has been obtained in animal experiments, most of it is derived from clinical observation, particularly in patients with pain originating in bone (metastases, Paget's disease, osteoporosis), but also in some cases of pain of extraskeletal origin (acute pancreatitis).

Animal evidence
Intracerebroventricular, but not subcutaneous, administration of 8 IU/kg salmon calcitonin raised the pain threshold in rabbits subjected to electrical stimulation of the dental pulp (upper incisors), one of the classic experimental models used to test for analgesic activity[202,212] (Fig. 79). This effect, which was rapid in onset, reached a peak at 90 minutes and was sustained. It did not involve opiate receptors[200] since it was not antagonized by naloxone or levallorphan[203,204], except at doses above those needed to reverse the effects of morphine. Furthermore, this antagonism is not wholly due to the hyperalgesic effect

Fig. 79   Effect of intracerebroventricular injection of salmon calcitonin (SCT) on pain threshold in rabbits subjected to electrical stimulation of the dental pulp[202]

of naloxone[204]. The analgesic activity of calcitonin appears to be sustained on repeated dosing[200], whereas the effect of equipotent doses of morphine (10 mg/kg) tends to decline with repeated administration. The effects of calcitonin and morphine together are additive, although they have been shown *in vitro* to act at different receptors.

Clinical evidence

A considerable volume of evidence has accumulated for the analgesic properties of calcitonin in man. At doses ranging from 50 to 400 IU/day, calcitonin has been found to relieve both bone pain (due to tumour metastases[205,207,210,213], Fig. 80, Paget's disease[198,211] and osteoporosis[205]) and pain not associated with disorders of bone, such as post-hysterectomy pain[199].

The route of administration is the most important factor affecting the time to onset of analgesic effect, being shortest after subarachnoid administration. The effective dose is also much lower by this route (for example, 4 IU salmon calcitonin compared with 12 IU by the peridural route via an indwelling catheter between L3 and L4) in patients with metastatic osteolysis[197].

The *type of calcitonin* is another important variable. Peridural injection of both 30 μg human calcitonin and 12 IU ($\simeq$ 2.5–3 μg) salmon calcitonin was followed by an analgesic effect within 20–30 minutes. However, the human calcitonin was fairly short-acting, whereas the effect of salmon calcitonin lasted 6–8 hours[197].

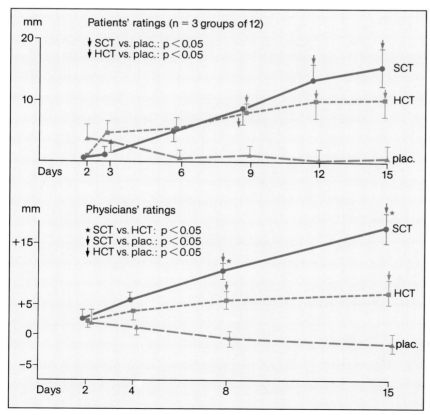

Fig. 80   Differences (mean ± SEM) between baseline pain and pain after treatment with SCT, HCT or placebo, as assessed using a vertical analogue scale, in patients with bone metastases[205]

The analgesic effect of calcitonin might be distinct from its anti-osteolytic effect since, whereas the hypocalcaemic effect after intravenous infusion using a minipump was not sustained for longer than 24–48 hours, the analgesic effect was maintained for some time longer ([211] and author's unpublished findings).

Possible mechanisms of the analgesic effect of calcitonin
A number of hypotheses have been advanced to explain the analgesic effect of calcitonin, although several of them probably reflect different aspects of a single mechanism:

– Increase in levels of circulating β-endorphin
Plasma levels of β-endorphin (but not of β-lipotrophin) were reported to be raised after intravenous administration of at least 50 IU salmon calcitonin in man[205,214] (Figs 81–83). Although these results have not been directly confirmed, a number of other findings support the 'endorphin hypothesis':

– Intracerebral administration of β-endorphin in rats and intravenous injection in man had a potent analgesic effect[215].

– The effects of β-endorphin, which is about 30 times more potent than morphine as an analgesic, were blocked by naloxone, a specific opioid antagonist[216].

– Some pain stimuli trigger β-endorphin release[217], with an increase in blood levels suggesting that the blood-brain barrier must be permeable to this peptide. Findings to this effect have also been made in rats[218].

– Inhibition of PGE_2 synthesis (Fig. 84)
This mechanism was proposed after calcitonin had been shown to possess anti-inflammatory activity in the carrageenan rat-paw oedema test[219]. It is supported by the finding that, when guinea-pig lung was perfused with a medium containing arachidonic acid

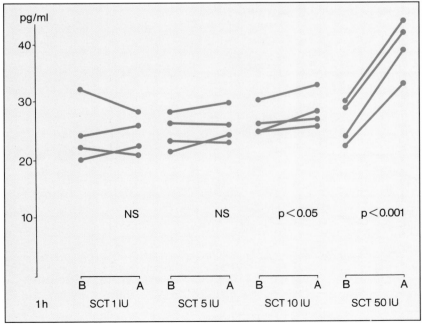

Fig. 81   Plasma levels of β-endorphin before (B) and after (A) i.v. injection of increasing doses of synthetic salmon calcitonin in patients with bone metastases[205]

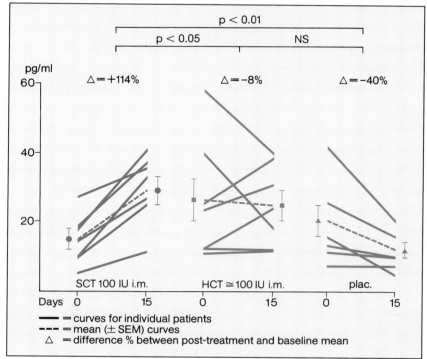

Fig. 82   Plasma concentrations of β-endorphin in 3 groups of 7 patients before and after treatment for 15 days with SCT, HCT or placebo[205]

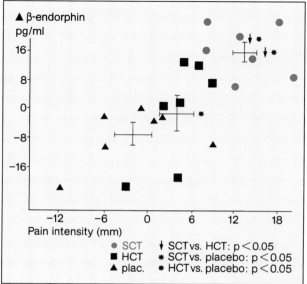

Fig. 83   Correlations between changes in plasma β-endorphin levels and in pain intensity ratings relative to baseline after SCT, HCT and placebo in patients with osteolytic metastases[205]

The large symbols represent individual values and the small symbols the group mean ± SEM.

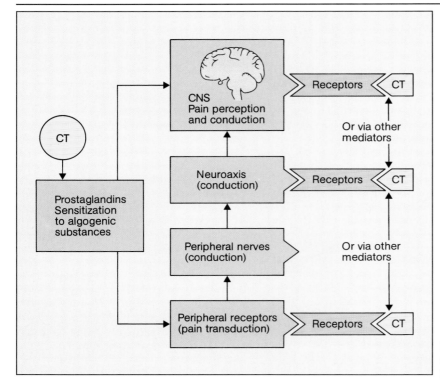

Fig. 84   Mechanisms possibly involved in calcitonin's inhibitory effect on pain via prostaglandin inhibition (based on [199])

in the presence of calcitonin, synthesis of prostaglandin E_2 and thromboxane A_2 was inhibited through a reduction in cyclogenase activity[220]. On the other hand, parenteral calcitonin did not reduce PGE_2 levels in man[221,222], and there has even been a report of hyperalgesia after prolonged administration of calcitonin[223]. Such cases are, however, very rare and are probably due to other pathological processes masking the analgesic effect of calcitonin.

– Interference with calcium flux
An effect on the movement of ionic calcium in neuronal membranes, particularly in the brain, is probably the principal mechanism of the analgesic effect of calcitonin[202]. This is suggested by the fact that the effect is antagonized by small intracerebroventricular doses

of calcium salts[224–226]. Calcium ion also antagonizes morphine-induced analgesia[227], and in mice subcutaneous injection of calcitonin (10 IU/kg on alternate days for 14 days) progressively raised the sensitivity threshold to the irritation induced by peritoneal injection of acetic acid[201,223].

– Involvement of the catecholaminergic system
The possibility that the catecholaminergic system might be involved in calcitonin-induced analgesia is suggested by the fact that the administration of 6-hydroxytryptamine significantly reduced the analgesic effect of salmon calcitonin as measured in rats by the hot-plate test[228]. Pretreatment with phentolamine or propranolol also significantly attenuated the effect of calcitonin but, when the two drugs were

administered alone, only propranolol had this effect. A more pronounced and sustained inhibitory effect on calcitonin-induced analgesia was exerted by atenolol.

– Involvement of serotonin

Another hypothesis, which also requires further clarification, is that the central serotoninergic system might be involved in the analgesic activity of calcitonin. It is based on the observation in male rats that the effects of intracerebroventricular calcitonin injection were antagonized by peripheral methysergide[229].

– Direct action on CNS receptors

Exogenous calcitonin is thought to cross the blood-brain barrier and to accumulate slowly in the brain, inducing analgesia once sufficient receptors are occupied[230]. This hypothesis is consistent with the clinical observation that the more peripheral the site of injection, the larger and more frequent are the doses in which calcitonin must be given in order to achieve and maintain analgesia. Accordingly, an analgesic effect is obtained with a smaller dose and more rapidly by the subarachnoid than by the subcutaneous route. Thus, whereas single subcutaneous doses of up to 50 IU/kg salmon calcitonin exerted a hypocalcaemic but not an analgesic effect, repeated subcutaneous doses induced a progressive rise in the threshold of sensitivity in the acetic-acid-induced abdominal-muscle-spasm model[201]. This effect began 12–48 hours after the last injection, was sustained and was not correlated with changes in plasma calcium. In the hotplate test, rats showed no significant rise in pain threshold until after the second or third injection of salmon calcitonin (40–80 IU twice daily for 3 days)[202,231].

– Indirect local effect

It is also possible that calcitonin reduces pain simply by effecting a local improvement in the vascular, nutritional, remodelling and inflammatory status of bone.

– Antidepressant effect

An antidepressant action of calcitonin, though not yet conclusively proved, may contribute to the hormone's analgesic effect[232,233].

● Neuromodulator effect

The role of calcitonin at sites other than bone has been discussed in several papers[195,197,230,234,235]. Both human and salmon calcitonin by intracerebroventricular injection potentiated haloperidol-induced catalepsy and inhibited apomorphine-induced hyperactivity in rats[197], particularly in aged animals[236]. It also affected levels of some neuromediators and their metabolites involved in the tuberoinfundibular system, anterior pituitary and extrapyramidal system (Table 46). It

Corpus striatum		Substantia nigra	
Dopamine	Ø	GAD	↓
DOPAC	Ø		
GAD	Ø		
Mediobasal hypothalamus		**Anterior pituitary**	
Dopamine	↓	Dopamine	↓
DOPAC	↓	DOPAC	↑

Table 46   Effect of calcitonin by intracerebroventricular injection on cerebral levels of dopamine, 3,4-dihydroxyphenylacetic acid (DOPAC) and glutamic acid decarboxylase (GAD) activity in rats[197]
↑ = increase   ↓ = decrease   Ø = no effect

caused dopamine depletion in dopaminergic nerve endings in the tuberoinfundibular system, increasing the availability of the neurotransmitter to the pituitary lactotrope cells and thus inhibiting prolactin secretion (Fig. 85). The raised pituitary level of 3,4-dihydroxyphenylacetic acid (DOPAC), the principal metabolite of dopamine, is also thought to result from the increase in the level of dopamine activity in the anterior lobe.

Conversely, the reduced level of the metabolite in the mediobasal hypothalamus may be explained by the low re-uptake affinity of dopamine in the median eminence, as a result of which it is rapidly lost into the portal system and therefore not metabolized[237,238]. Inhibition of prolactin secretion would thus appear to be mediated by direct interaction of calcitonin with the dopaminergic system. A less likely alternative is that calcitonin lowers plasma levels of prolactin by in-

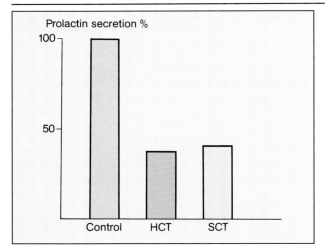

Fig. 85   Effect on plasma prolactin levels in rats of intracerebro-ventricular administration of equivalent doses of two different calcitonins[197]

4–8 hours after administration by the peripheral routes. An anorectic effect was also observed in a toxicological study in rabbits (author's unpublished findings). It is not known whether this effect is connected with inhibition of gastric secretion or whether it is an independent effect of central origin[171].

Summary: CNS effects of calcitonin

Analgesia is the principal central effect of calcitonin and has been reported in many studies in both animals and man. The many hypotheses concerning the mechanism of this action include an effect on calcium flux in the neuronal membrane, an action at specific central receptors, an increase in β-endorphin levels, inhibition of prostaglandin synthesis, or simply an in-

creasing the rate at which it is metabolized at the periphery[235].

At the level of the extrapyramidal motor system the action of calcitonin is not thought to be due to an effect on dopaminergic pathways. This is supported by the fact that it does not affect the dopamine or DOPAC content of the corpus striatum. It does, however, reduce glutamic acid decarboxylase (GAD) activity in the substantia nigra, indicating inhibition of the synthesis of γ-aminobutyric acid (GABA) by the GABAergic fibres of this nucleus. These observations suggest that the effects of calcitonin by intracerebroventricular injection on the behaviour of rats may be due to modification of nigrostriatal GABAergic transmission[194,195,239].

● Anorectic effect in animals

In rats and rhesus monkeys given calcitonin systemically or by direct intracerebroventricular injection, food intake was reduced and weight gain retarded[194–196,240,241]. In rats, peak anorectic effect was observed

1.	**Endocrine interactions**
	● With calcium-regulating hormones
	● With hormones of the gastrointestinal tract
	● With anterior pituitary hormones
	● With β-endorphin
2.	**Anti-inflammatory effect**
	● In experimental models
	● In acute pancreatitis
3.	**Cardiovascular effects**
	● Flush – control of blood flow: in bone? renal?
	● Prevention or slowing of the immunoarteriosclerotic process?
4.	**Antistress effect**
	● Hypothetical (mediated by inhibition of catecholamine secretion?)
5.	**Other effects**
	● Reduced synthesis of prostaglandins (isolated organ)
	● Effect on 'plasminogen activator'
	● Activation, followed by inhibition, (in vitro) of multiplication of tumour cell lines (breast cancer)?
	● Inhibition of the secretion of salivary (?) and serum amylase (in acute pancreatitis)

Table 47   Other confirmed, probable or postulated effects of calcitonin

direct result of a general improvement in all aspects of painful bone lesions resulting in reduced pain perception at central nervous level. The analgesic effect of calcitonin probably depends on a combination of several factors, the foremost being changes in calcium flux.

It is also fairly well established that the more peripheral the site of calcitonin administration, the longer is the delay and the larger the dose required to produce an analgesic effect. The various routes of administration can be placed in order of decreasing efficacy against bone pain as follows:
intraventricular/intrathecal > peridural/epidural > intravenous > intramuscular > subcutaneous

Other putative central effects or mechanisms such as appetite suppression are unconfirmed and/or require much more detailed study. However, all the evidence suggests that calcitonin has a neuromodulator function in the central nervous system[144].

Other effects (Table 47)

As research continues, new properties of calcitonin are continually being discovered. Some of these may turn out to have therapeutic application, but their immediate interest lies in what they reveal about the mechanisms of action of calcitonin and the pathophysiology of certain bone and other diseases.

Hormone	Species	Effect	Comments	References
PTH	Man, rat	↑	In response to hypocalcaemia	8.69.97.242−245
1,25 (OH)$_2$D$_3$	Man, rat	↑	Via 1α-hydroxylase	119.120
Gastrin	Man, rat	↓	Fasting, after food. Healthy subjects and duodenal ulcer patients	246−248
Insulin	Man	↓	Transient, sometimes at the beginning of treatment, sometimes in response to glucose Arginine-stimulated increase	248−250
	Man	∅	Long-term treatment	146
VIP	Man	↑	HCT>SCT; causes flushing	251
Somatostatin	Man	↑	The raised somatostatin level may affect other hormones of the gastrointestinal tract.	180.252
TSH	Man, rat	↓	In some cases; basal or stimulated No change acc. to [256]	255
LH	Man, rat	↓	Basal or stimulated No change acc. to [256]	255
GH	Man	↓	In some cases No change acc. to [256]	149.248
ACTH	Rat	↓	In plasma following 100 IU/kg i.p.	258
	Rat	↑	In the isolated hypothalamus and pituitary	258
	Rat	↑	Normal and stressed/i.c.v. injection	257
	Man	↑	100 IU SCT i.v.	214
PRL	Man, rat	↓		See Table 49
β-endorphin	Man	↑		205.214.271
Prostaglandins	Guinea-pig	↓	Isolated organ (lung) In inflammatory processes?	220.221 222

Table 48   Effect of calcitonin administration on the production of various hormonal and other factors
↑ = rise   ↓ = fall   ∅ = no change

– Interactions with other endogenous hormones (Table 48)

Hormones involved in calcium metabolism

– Parathyroid hormone
Calcitonin administration is followed by an almost immediate rise in the blood level of PTH[242], its functional antagonist in the regulation of calcium metabolism, although the interaction between the two hormones is in fact indirect, being mediated by the blood level of calcium. The indirect nature of their interaction is underlined by the fact that the activation of adenylate cyclase in bone and kidney cells by PTH is not blocked by calcitonin, that the parathyroid glands continue to function normally during long-term treatment with calcitonin[243,244], that hyperparathyroidism is not typical in medullary carcinoma of the thyroid[245] and, above all, by the fact that the two hormones act at different receptors in bone and kidney[8,69,97].

– $1,25(OH)_2D_3$
As we have already seen, calcitonin stimulates 1-α-hydroxylase activity, thereby increasing production of $1,25(OH)_2D_3$, the active metabolite of vitamin D[119,120].

Hormones of the gut and pancreas
(See also "Effects on the GI tract and accessory organs" above.) Although calcitonin inhibits secretion of most of these hormones, like gastrin[246–248] and insulin[248–250], it is thought to increase production of *vasoactive intestinal polypeptide* (VIP) (Fig. 86), which may be the cause of the flushing often experienced as a side effect of treatment with calcitonin, especially human calcitonin[180] (Fig. 87). Calcitonin also appears to stimulate the production of *somatostatin*[251,252], which was first isolated from the hypothalamus in 1973. Somatostatin is also secreted by the D cells of both the pancreas and the gastrointestinal tract and has been found to display a similar pattern to that of calcitonin with regard to its interactions with gastrointestinal hormones. Conversely somatostatin seems

to inhibit the secretion of calcitonin in both man and animals (as reviewed in [10]).

Anterior pituitary hormones
Most of the work done on the effect of calcitonin on the hormones of the anterior pituitary has been concerned with prolactin[235,253]. However, other hormones are also affected[144,145,254], notably *thyroid-stimulating hormone* (TSH)[256], *luteinizing hormone* (LH)[255] and *growth hormone* (GH)[149,248], secretion of all of which seems to be slightly inhibited by calcitonin in man (although not according to [256]).

In rats intraperitoneal doses of 100 IU/kg salmon calcitonin have been reported to lower the plasma level of *adrenocorticotrophic hormone* (ACTH) but to increase its concentration in the hypothalamus and, to a lesser extent, in the mesencephalic and cortical areas of the brain[257]. However, very high doses of calcitonin were used in this work. *In vitro*, release of ACTH from the isolated rat pituitary was enhanced

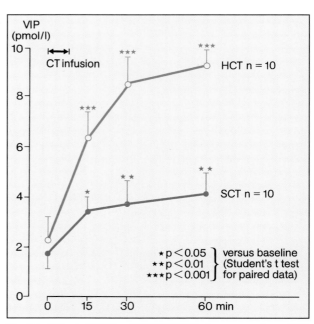

Fig. 86   Effect of HCT and SCT (100 IU i.v.) on mean plasma VIP in healthy volunteers[251]

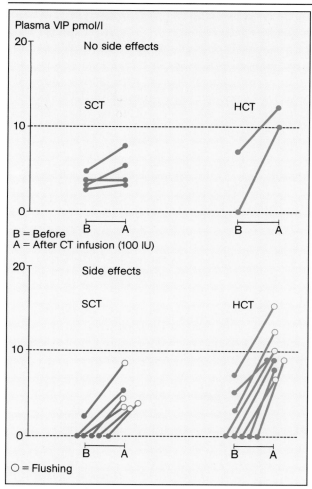

Fig. 87   Relationships between plasma VIP and side effects in 10 patients given infusions of SCT or HCT. The open circles denote those experiencing hot flushes[251].

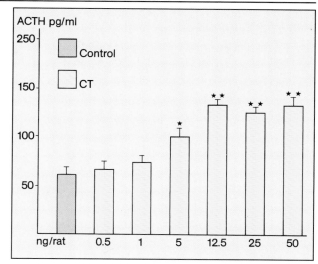

Fig. 88   Effect in rats of various doses of SCT on serum ACTH levels 30 min after intracerebroventricular injection[258]

Values are means ± SEM of 6 rats per group
* = $p < 0.05$, ** = $p < 0.01$ versus controls

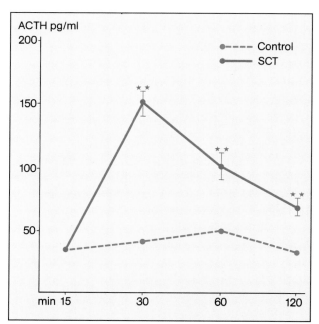

Fig. 89   Serum ACTH levels at various times following SCT 50 ng/rat by intracerebroventricular injection[258]

Values are means ± SEM of 6 rats per group
** = $p < 0.01$ versus controls

in the presence of calcitonin (50 IU/ml)[144,257,258], while serum ACTH was significantly increased by salmon calcitonin in rats given intracerebroventricular doses of 5 ng ($p < 0.05$) and 12.5 ng, 25 ng and 50 ng ($p < 0.01$)[257] (Fig. 88). This effect was maximal at 30 minutes and still significant at 60 min and 120 min (Fig. 89). The highest dose level also produced a significant ($p < 0.05$) increase in serum ACTH concentration in stressed rats (Fig. 90).

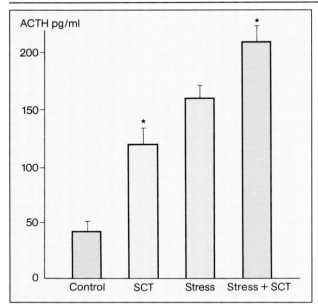

Fig. 90   Serum ACTH levels in stressed and unstressed rats given SCT 50 ng by intracerebroventricular injection[258]

Values are means ± SEM of 6 rats per group
* = p < 0.05 versus controls

Stimulation of ACTH secretion by calcitonin might be effected through central noradrenaline and/or serotonin pathways that regulate corticotrophin-releasing-factor activity[257,359]. It has also been reported that the response of both healthy volunteers and of patients experiencing acute episodes of osteoporosis is characterized by a significant increase in ACTH levels, as well as in those of *β-endorphin* and *cortisol*[214].

While the published findings on the effect of calcitonin on *prolactin* release and blood levels are far from consistent (Table 49), the overall picture is one of inhibition at normal dose levels and by the usual routes of administration (i.v., i.m. and s.c.); only intracerebroventricular (i.c.v.) injection, usually of high doses, has been associated with stimulation of prolactin secretion. Details of the principal relevant findings in man and rats are as follows:

In humans the release of prolactin is reported to be inhibited by intravenous[235,253,256] and subcutaneous[259] calcitonin (salmon in all cases), although other authors have reported little or no effect for the subcutaneous[260] and intramuscular[248] routes. Rats likewise normally appear to respond to calcitonin with a decline in prolactin secretion[193,235,261,262], including where this has

in vitro	No effect on rat anterior pituitary tissue exposed to ECT[263]
	Slight stimulatory effect on cultured rat anterior pituitary cells perfused with ECT[265]
in vivo	
Rat	Depression of basal blood levels and suppression of peaks induced by morphine, heat stress and suckling in lactating rats given i.v. or i.c.v. injections of SCT[193,235,261-263]
	Clearcut decline in basal plasma levels after 2.5–10 µg/kg i.v. and 2.5–25 ng/animal i.c.v. SCT[193]
	Slight decrease in basal plasma levels in conscious male rats after i.v. injection of ECT[263]
	Inhibition of physiological secretion by "supra-physiological doses" of SCT i.c.v.[235]
	No effect after i.v. injection[263,264], but stimulation in anaesthetized (urethane) male rats given ECT, HCT or SCT by i.c.v. injection[263-265]
	Transient increase followed by a significant decrease in plasma levels in conscious male rats given high i.c.v. doses (2.5 µg) of ECT, compared with only a fall after low doses (2.5 ng)[263]
	Rise in suckling-induced secretion after ECT by i.c.v. injection[261]
Man	Little or no increase of basal serum levels in healthy volunteers after s.c. injection of HCT[260] or i.m. injection of SCT[248]
	Decrease in serum levels in psychiatric patients after s.c. injection of SCT[259], in plasma levels in hyperprolactinaemic patients after SCT i.v.[256] and in healthy volunteers treated i.v.[253,256]
	Depression of basal blood levels and inhibition of the response to TRH after i.v. infusion of SCT[235,256]

Table 49   Effect of calcitonin on prolactin release (principal published findings)

been artificially stimulated by stress or morphine[193]. On the other hand some reports suggest that, in anaesthetized animals at least, intravenous calcitonin has no effect on prolactin secretion[263-265] and that intracerebroventricular administration even causes a significant, dose-related increase[264,265]. In one report an initial increase is described in conscious rats given high intracerebroventricular doses, followed by a significant fall in blood levels of the hormone[263].

The inhibitory effect of calcitonin on prolactin secretion is independent of its effect on extracellular calcium concentration[235,253,259], being mediated via a direct action on the dopaminergic system[264], although high doses may have a qualitatively similar effect on the pituitary[235]. This is suggested by the fact that its effect on prolactin is abolished by treatments interfering with dopaminergic neuron function (e.g. haloperidol[266]) and by surgical interruption of the tuberoinfundibular dopaminergic pathways at the level of the median eminence, prolactin secretion being regulated by dopamine released from this system[267,268]. The hypothalamus also contains the highest concentration of calcitonin receptors.

In patients with a prolactinoma, basal plasma prolactin levels ranging from 42 ng/ml to 4130 ng/ml did not change in response to intravenous infusion of Asu[1,7]-eel calcitonin (ECT), nor did ECT(10^{-9}–10^{-6} mol/l) affect prolactin release from prolactinoma tissue perfused *in vitro*. These findings suggest that ECT at least may not act directly on the pituitary to modify prolactin release. On the contrary, peripherally administered ECT appears to suppress prolactin release via the central nervous system[269].

Evidence in rats that intracerebroventricular calcitonin does not reduce serotonin synthesis or release suggests that it might interfere with polysynaptic mechanisms involved in the facilitation of prolactin secretion by serotoninergic neurons[270].

β-endorphin
Central and peripheral levels of β-endorphin are reported to be raised after administration of calcitonin[205,214], producing an analgesic effect. For further details, see above.

● Anti-inflammatory properties

Calcitonin exhibits anti-inflammatory properties in the following models of acute or chronic inflammation in rats and mice[219,271,272]:
– histamine oedema (mice)
– dextran oedema (rats)
– nystatin oedema (rats)
– carrageenan oedema (rats)
– yeast hyperthermia (rats)
– acetic-acid-induced abdominal constriction (mice) (pain due to inflammation)
– Freund's-adjuvant arthritis (rats)

The anti-inflammatory effect is dose-dependent and, in acute experiments[272] at least, involves inhibition of the first phase of the inflammatory process; no antipyretic effect is exerted. In long-term experiments (Freund's-adjuvant arthritis) salmon calcitonin reduced oedema and the response to pain stimuli, restricted the number and severity of articular lesions and consequently improved the range of movement in the joints affected by the arthritic process[272]. In carrageenan oedema the potency of salmon calcitonin is three times greater than that of human or porcine calcitonin (at doses expressed in units of biological activity)[219].

The mechanism of action probably involves a reduction in vascular permeability[273] and stimulation of lysine decarboxylase activity, as in the case of many of the non-steroidal anti-inflammatory agents. This effect is reported to be independent of any changes in tissue calcium levels[144]. In addition, dose-dependent inhibition of prostaglandin and thromboxane biosynthesis is reported to occur as a result of partial inhibition of the activity of cyclo-oxygenase, the enzyme responsible for the first stage of arachidonic-acid metabolism[220,273]. However, this effect has not yet been observed in man; in patients with hypercalcaemia associated with a tumour and with raised prostaglandin levels, calcitonin exerted both anti-hypercalcaemic and analgesic effects but had no effect on prostaglandin levels[274] (Fig. 91).

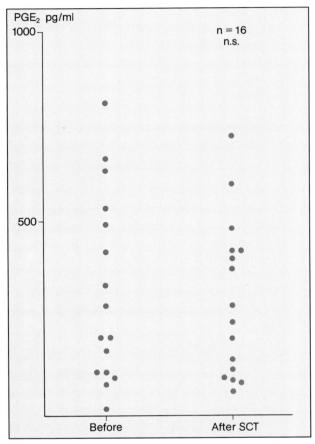

Fig. 91   SCT 100 IU daily for 30 days had no significant effect on plasma PGE$_2$ levels in 16 patients with neoplastic hyper-calcaemia[274].

● Cardiovascular effects

Although calcitonin is thought to have little or no effect on the cardiovascular system, a number of observations are of interest:

One of the commonest side effects of calcitonin is flushing. This phenomenon, which occurs in both normal states such as emotion and menopause, and pathological conditions such as the 'multiple paraneoplastic diseases', is thought to result from a peripheral effect, possibly due to the release of various factors, including the vasodilator vasoactive intestinal polypeptide.

This side effect is sometimes associated with the peak plasma level of calcitonin, especially in the early stages of treatment and following intravenous administration.

It is possible that calcitonin is a vasoactive hormone[275] and that its beneficial effect on the type of bone pain sometimes called 'migraine of the bone' may be due to restoration of normal bone blood flow. Also, its diuretic and natriuretic effects might be the result of a substantial increase in renal blood flow, supporting the hypothesis that calcitonin is a renal vasoactive peptide[276].

In rats, very high doses of calcitonin (SCT 100 IU/kg) have been reported[144] to inhibit the arrhythmic response to aconitine and ouabain, but not to adrenaline.

Intravenous salmon calcitonin produced a pressor response in rats made hypotensive by bleeding but was without effect in normotensive rats or those made hypotensive by pithing. The pressor effect was significantly attenuated by chemical sympathectomy, suggesting that the hormone potentiates sympathetic outflow. In both normotensive and bled rats, centrally administered SCT produced a pressor response which was not significantly attenuated by chemical sympathectomy (C.A. McArdle, personal communication).

Long-term administration of porcine calcitonin reduced the extent of the arteriosclerotic process in an experimental model (immuno-arteriosclerosis) in rabbits[277-279], presumably by inhibiting the formation of calcareous deposits in arterial tissue by preventing the entry of calcium ion into smooth muscle.

● Anti-stress effect

The suggestion that calcitonin might have a protective action against stress[280] is supported by the following facts:

Stress is invariably associated with stimulation of the autonomic nervous system, and particularly of catecholamine secretion. This stimulates osteolysis, causing hypercalcaemia[281,282] which, in turn, increases the secretion of calcitonin.

Thus, many stressful situations are associated with an increase in the blood calcitonin level. It is also known that even normal concentrations of catecholamines tend to increase calcitonin secretion in sheep[283], hens[284] and rats[282].

The autonomic nervous system and calcitonin secretion are thus closely linked, and certainly the hypothalamus, the control centre of the autonomic nervous system, has a high density of calcitonin receptors. Theoretically, it should be possible to intervene in this 'stress cascade' by administering calcitonin at a suitable dosage[232].

Calcitonin, at least in animals, has an anti-ulcerogenic effect which is independent of its hypocalcaemic effect and attributed in part to a central mechanism[171]. This could also be regarded as an anti-stress effect.

These findings, and others, combine to suggest that calcitonin may have a beneficial effect on the autoexcitation process[285]. If this is so and calcitonin does exert a sedative effect in certain circumstances, it might be possible to use it as a means of reducing doses of psychotropic drugs. The specific use of calcitonin in the exacerbation phase of agitated psychoses and in severe depression has been reported[233,286].

● Miscellaneous effects

A number of other, apparently unrelated, effects of calcitonin have been described:

Reduced prostaglandin synthesis[220-222]
Stimulation of plasminogen activator[286]
Inhibition, following initial stimulation, of growth in cultures of certain cell lines derived from breast tumours[287,288]
Potentiation of some of the effects of kanamycin in rats (effects on blood pressure, heart and respiration rates, head drop, and loss of corneal reflex)[289]
Stimulation of amylase secretion by salivary glands in man[290]

These less well-known effects of calcitonin give some idea of the many aspects of this hormone, notably those involving the highly complex endocrine system, which still require detailed elucidation. They are secondary to its primary, anti-osteoclastic action, and some are seen only at very high doses or appear to have no practical therapeutic potential. This applies, for example, to the anti-ulcerogenic and anti-inflammatory effects.

Other properties, however, either have direct clinical relevance (analgesic effect in bone metastases) or else help to explain side effects like flushing and diarrhoea. The 'calcium connection', i.e. the involvement of calcium in its biological role as second messenger at cell, organ, system or whole-body level, is common to almost all of these properties, although one or two of them – for example the effect on CNS neurotransmission and prolactin secretion – might possibly be independent of the intracellular and extracellular movement of calcium.

The biopharmaceutics of calcitonin

Methods of assay and routes of administration

These two factors have a major influence on biopharmaceutical data and should always be borne in mind, especially when comparing different calcitonin preparations.

● Assay methods

The results of measurements of the concentrations and clearance of injected calcitonins in blood and urine will depend on the experimental method used. The two principal methods are measurement of radio-

activity after injection of a tracer dose of calcitonin labelled with radioactive iodine (^{125}I or ^{131}I) and measurement by a radioimmunoassay (RIA) specific for the type of calcitonin injected. As the measurement of radioactivity does not confirm that the radioactive label is still attached to the calcitonin peptide, RIA is the technique most often used. However, RIA measures the concentration of immunoreactive calcitonin without or following extraction, which is not necessarily the same as the concentration of biologically active calcitonin. The sensitivity of the assay method (limit of detection) is obviously an important factor because of the relatively low concentrations, depending on the calcitonin species (e.g. SCT or ECT), at which the hormone is biologically active.

Author	Subjects	n	Route	Dose	Assay method	Metabolic clearance rate (ml/min)	Half-life of elimination (min)	Distribution volume (litres)
Beveridge[295†]	Patients	6 6 6	i.v. i.m. s.c.	35 μg	RIA	477	11.6/67.8 58.2 88.2	11.1
Nüesch[291]	Healthy volunteers	16	s.c.	19.9 μg	RIA	212	87.0	20.3
Huwyler[296]	Healthy volunteers	6	i.v. infusion	0.17 μg/min[‡] over 240 min	RIA	255	14.8	4.8[§]
Singer[298]	Patients	2	i.v. infusion		RIA	261	12.9/51.6	
Ardaillou[299]	Healthy volunteers	8	i.v.	0.4 μg	^{125}I-SCT	184		3.5

Table 50   Pharmacokinetic data for salmon calcitonin

† Other relevant data from this study:
  Absorption half-life = 23.4 min
  MCR value was 3–5 times lower in a patient with renal insufficiency.
  $C_{p\,max}$ = 447 ± 37 pg/ml (384 ± 33 pg/ml when calculated from the mean concentration)
  t_{max} = 49.2 ± 7 min

Plasma protein binding = 30–40%
Absolute bioavailability = 71% (s.c. route) and 66% (i.m. route) respectively
Urinary excretion = < 2% of dose
‡ 0.23 μg/min in one subject
§ Calculated from published data

Author	Subjects	n	Route	Dose	Assay method	Metabolic clearance rate (ml/min)	Half-life of elimination (min)	Distribution volume (litres)
Huwyler[296]	Healthy volunteers	8	i.v. infusion	0.17 μg/min[†] over 240 min (40 μg in total)	RIA	556	9.1 (n=7)	7.2‡ (n=7)
Bijvoet[300]	Patients	3	i.v.	75 μg	RIA	708	4.2/16.2	5.5‡
Singer[298]	Patients	2	i.v. infusion		RIA	661	7.4/63.5	
Ardaillou[299]	Healthy volunteers	8	i.v.	0.4 μg ^{131}I-HCT		144		2.8
Ardaillou[292]	Healthy volunteers	10	i.v.	0.16−0.82 μg ^{125}I-HCT		110	2.8/11.5/ 318	4.5
	Patients (with renal insufficiency)	10	i.v.	0.16−0.82 μg ^{125}I-HCT		21.6	4.5/22.5/ 1564	5.3
	Healthy volunteers	3	i.v. infusion	0.03−0.15 μg/h ‡ ^{125}I-HCT[§]		148‡		
	Patients (with renal insufficiency)	3	i.v. infusion	0.03−0.15 μg/h ‡ ^{125}I-HCT[§]		51 ‡		

Table 51   Pharmacokinetic data for human calcitonin

† Higher in 3 subjects   ‡ Values based on data available   § Total 0.24−1.2 mg

Author	Subjects	n	Route	Dose	Assay method	Metabolic clearance rate (ml/min)	Half-life of elimination (min)	Distribution volume (litres)
Riggs[302]	Healthy volunteers	5	i.v. infusion	2 mg/h over 3 h	RIA	846	3.2/70.7	12.6
	Healthy volunteers	5	i.v.	1 mg			2.2/27.3	
	Osteoporotic patients	6	i.v.	1 mg			2.4/55.7	
	Pagetic patients	5	i.v.	1 mg			2.4/19.8	
	Patients with hyperparathyroidism	5	i.v.	1 mg			1.5/21.0	

Table 52   Pharmacokinetic data for porcine calcitonin

● Routes of administration

The most important point relevant to the biopharmaceutics of the calcitonins is that at present they can be administered only parenterally, the oral route being impossible because of degradation by aminopeptidases or proteases in the gastrointestinal tract. However, research is proceeding in several areas with the object of developing non-injectable dosage forms (e.g. intranasal, intrarectal), particularly for patients requiring long-term treatment.

Pharmacokinetics

Tables 50–52 show reported values for some of the principal pharmacokinetic variables for the main calcitonins in therapeutic use. Of these, synthetic salmon calcitonin is the most extensively studied and eel calcitonin (in most cases the [Asu^{1,7}] analogue), being the most recent, the least. Urinary excretion may be ignored since radioimmunoassay has shown that less than 2% of the dose is excreted by this route. However, 95% of the radioactivity from a dose of radio-labelled calcitonin was recovered within 48 hours, although unchanged drug apparently accounted for only 2.4% [291,292].

● Synthetic salmon calcitonin (Table 50)

In a study[291] in 16 healthy volunteers, a dose of 19.9 μg synthetic salmon calcitonin (SCT) administered subcutaneously was found to be rapidly absorbed, with a mean (± SEM) half-life of 23.4 (± 4.2) min. The mean peak plasma concentration was 384 (± 33) pg/ml (mean of individual peaks = 447 [± 37] pg/ml) and was reached at a time (t_{max}) 1 h after administration (mean of individual t_{max} values = 49 [± 7] min; Fig. 92). Concentrations were below the detection limit of the immunoassay 12 h after administration. The apparent volume of distribution after this subcutaneous dose was 20.3 (± 1.9) litres, compared

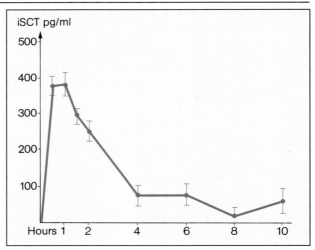

Fig. 92   Mean (± SEM) SCT plasma level curve for 16 healthy volunteers given a dose of 19.9 μg s.c.[291]

with 28.6 (± 2.7) litres assuming an absorption of 100%, and with a reported range of 3.5–11.1 litres. Plasma protein binding amounted to 30–40%; according to the literature, calcitonin in human blood is bound to specific carrier proteins – IgM, α_2-macroglobulin, α_2-lipoprotein and albumin in decreasing order of affinity[186,293]. In rats the principal protein binding SCT is albumin[294].

The elimination half-life in this study was 87 (± 18) min and the uncorrected metabolic clearance rate (f = 1) was 299 (± 30) ml/min, compared with 212 (± 21) ml/min after correction for the subcutaneous route. Reported values for the metabolic clearance rate of SCT range from 184 to 477 ml/min. In two studies[295,296], elimination after intravenous injection was found to be biphasic, with values of 11.6 and 12.9 min respectively for the shorter and 67.8 and 51.6 min for the longer phase. After subcutaneous and intramuscular injection, only the longer half-life of elimination was observed, with values of 88.2 and 58.2 min respectively; these are consistent with the half-lives reported for intravenous injection. The data thus do not support the contention that the half-life is longer after intramuscular than after intravenous injection[297], although this claim is based on pharmacodynamic studies involving variables not comparable with those obtained from the plasma data. In one study[296] in which SCT was given by infusion at a rate of 0.17 μg/min for 240 minutes a very short half-life of elimination was found.

● Synthetic human calcitonin (Table 51)

Reported values for the metabolic clearance rate of synthetic human calcitonin (HCT) after intravenous infusion or injection range from 110 to 708 ml/min, depending on the method used. The latter figure was obtained by radioimmunoassay in 3 patients given 75 μg by intravenous administration[300], whereas a mean metabolic clearance rate of 556 ml/min was reported for 8 healthy volunteers after infusion at 0.17 μg/min for 240 min[296]. [125]I and ^{131}I-HCT, on the other hand, gave values of only 110–148 ml/min[292,299]. In patients with renal insufficiency, MCR values were three to five times lower than those in normal subjects[292], indicating that blood levels are higher in these patients and that the dosages required are therefore lower.

One, two or even three half-lives of elimination have been derived for HCT, depending on the method of kinetic analysis. The shortest were of 2.8, 4.2 and 4.5 min, with an intermediate phase of 9.1–63.5 min. A third phase of 318 min was observed in one case, but only with a method using radiolabelled hormone. In patients with renal insufficiency, the short and medium half-lives were almost doubled and the long half-life was about five times longer (1564 min) than in normal subjects[292].

Reported values for distribution volume range from 2.8 to 7.2 litres.

● Natural porcine calcitonin (Table 52)

The pharmacokinetic characteristics of natural porcine calcitonin (PCT), as studied in human subjects given ^{125}I-PCT intravenously (Fig. 93), are as follows:

$T_{1/2}$ of elimination	3 min 30 s (α phase)
	1 h (β phase)
Distribution volume	2.82 litres
Metabolic clearance rate	5.5 litres.h^{-1}
Plasma clearance	see Fig. 93

With radioimmunoassay, similar half-lives of elimination (1.5–2.4 min) were found in healthy volunteers and osteoporotic, pagetic and hyperparathyroid patients given intravenous doses of 1 mg PCT[302]. Long half-lives ranged from 19.8 to 55.7 min, the highest

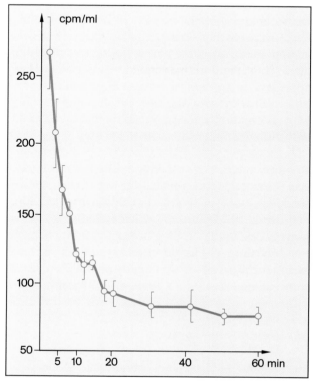

Fig. 93   Plasma clearance curve for porcine calcitonin administered by i.v. injection to human subjects[186,301]

values occurring in the osteoporotic patients. Calculations based on data obtained after administration of PCT by infusion at about 2 mg/h for 3 h to five healthy volunteers gave the following values: metabolic clearance rate, 846 ml/min; distribution volume, 12.6 litres; short and long elimination half-lives, 3.2 and 70.7 min respectively.

● Aminosuberic[1,7] analogue of eel calcitonin (ECT)

In rats given ^{125}I-ECT by intramuscular injection, peak plasma concentration was reached after 10–30 min. Significant levels of radioactivity were also found in the kidneys, stomach and liver of the animals, reaching maximum values after 30–120 min and falling thereafter. About 42% of the radioactivity contained in the dose was excreted in the urine within 4 h and 74% within 48 h. Only 7% was recovered in the faeces over a 48-hour period. In man, peak blood levels were reached 30 min after an intramuscular injection and levels were still measurable at 120 min. These data are as given in Japanese promotional literature and do not appear to have been formally published. Strictly speaking they need complementing with information about the dose administered and the detection limit of the assay methods used.

In other studies (also unpublished) plasma curves of Asu[1,7] eel calcitonin in healthy volunteers given a single i.v. dose of 40 I.U. indicated an elimination half-life of 10.34 minutes, which is similar to that of salmon calcitonin. In nine healthy volunteers given the same dose by i.m. injection, a peak plasma level of about 300 pg/ml was apparently attained and levels were still high about four hours after administration.

Metabolic clearance rates

Here again, published reports are far from unanimous. The metabolic clearance rate (MCR) of human calcitonin in man, for example, was reported to be

MRC (ml/min)	Type of calcitonin	Method
846	PCT	RIA
708	HCT	RIA
661	HCT	RIA
556	HCT	RIA
477	SCT	RIA
261	SCT	RIA
225	SCT	RIA
212	SCT	RIA
184	SCT	Radiolabelling
144	HCT	Radiolabelling
110	HCT	Radiolabelling

Table 53   Metabolic clearance rates for various calcitonins in descending order[291]

about twice that of salmon calcitonin[296,298], but other reported values were greater for salmon than human calcitonin[299]. However, this discrepancy might be explained by the use of radioimmunoassay in one study[296] and of radiolabelling in the other[298].

Table 53 presents MCR values for HCT, PCT and SCT; no values appear to have been published yet for ECT. Apart from those for radiolabelled HCT, the values present a coherent overall picture; they are highest for PCT, followed by HCT and then by SCT. The low MCR values found with radiolabelled calcitonin might be explained by the tendency of this method to overestimate blood levels and the fact that "radioactive calcitonin" (i.e. radioactivity assumed to represent calcitonin) is measurable for much longer than immunoreactive calcitonin. Activity/time curves exhibiting up to three half-lives have been described[292], the longest of them extending over several hours. This results in a high value for the area under the activity/time curve and a low value for MCR. Furthermore, only a tiny fraction of the radioactivity recovered in the urine is accounted for by unchanged calcitonin[292,296]. Consequently, values obtained with radioimmunoassay probably underestimate the time values, whereas those derived from radioactivity measurements overestimate them. The following mean MCR values are based on results obtained with radioimmunoassay only[291]:

	SCT		HCT		PCT
MCR	271*	<	607*	<	846
Ratio	1.0	:	2.2	:	3.1
			1.0	:	1.4
(*Weighted mean)					

The metabolic clearance rate of HCT thus appears to be twice as high as that of SCT and the value for PCT three times as high. This accords with the report[296] that the mean MCR value for HCT is 2.7 times higher than that for SCT (Fig. 94). The lower rate of metabolic clearance is part of the explanation for the fact that the hypocalcaemic potency of salmon calcitonin – like that of ECT[184] – is about 30 times greater than that of the human and porcine types (weight for weight).

Plasma levels in relation to hypocalcaemic effect

No direct correlation has yet been conclusively demonstrated between the kinetics of calcitonin in plasma and its hypocalcaemic effect, though this may be due to the limited sensitivity and variable performance of the analytical methods used. However, the many investigations carried out in man and animals show that the effect is dependent on the dose and the route of administration. After intravenous injection the hypocalcaemic effect is apparent almost immediately but there is a lag after intramuscular and subcutaneous injection. Maximum hypocalcaemic effect is attained rather later than peak plasma level: in pagetic patients receiving 80 IU SCT or 0.5 mg HCT by the intramuscular route the maximum effect was observed after more than 4 hours, while in another study in which 100 IU HCT were administered subcutaneously to pagetic patients the peak effect occurred almost 6 hours later than the peak plasma level[111] (Fig. 95). In general, the duration of hypocalcaemic effect is longer (8–15 or even 24 h, depending on the dose) than the period during which calcitonin is detectable in plasma.

The kinetic profile of the hypocalcaemic effect of ECT in man has been found on the whole to be the same as that of a biologically equivalent dose of SCT[184], and the relation between biological effect and plasma level

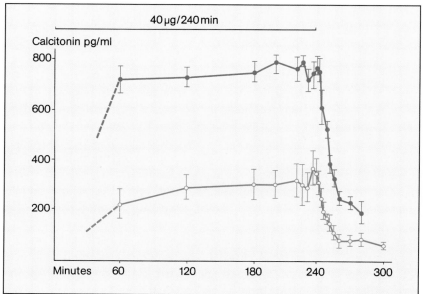

Fig. 94   Mean (± SEM) plasma profiles for 5 healthy volunteers given 40 μg HCT(o—o) or SCT (●—●) by infusion over 240 minutes, as measured by RIA techniques for 1–32 homologous HCT and SCT[296]

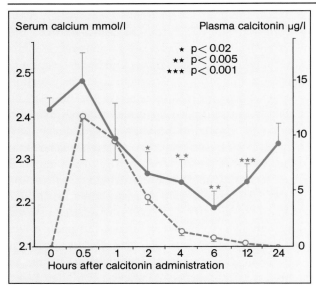

Fig. 95   Serum calcium (——) and plasma calcitonin (----) levels (mean ± SEM) in 7 pagetic patients after a single s.c. dose of 100 IU synthetic HCT[111]

of immunoreactive hormone is therefore probably similar for these two types. In the ionized calcium kinetics model in young rabbits, however, the hypocalcaemic potency of SCT appeared to be approximately 20% greater than that of ECT[303] (see Fig. 108).

Pharmacokinetic data as a guide to dosage

As with most drugs, pharmacokinetic data for calcitonin obtained by radioimmunoassay and those obtained by radiolabelling do not coincide; for example, RIA tends to underestimate and radiolabelling to overestimate the half-life. This means that the monitoring of plasma drug levels by these methods is not a reliable way of ascertaining the dosage requirement of the individual patient, especially in cases of severe hypercalcaemia. Dose-finding must therefore be based on biochemical or, even better, clinical variables – at least until more sensitive assay techniques

are developed and/or a clear relation is found between blood level and effect.

Bioavailability

The absolute bioavailability of SCT was reported[295] as 66% for the intramuscular route and 71% for the subcutaneous route after a dose of 35 μg. In fact, calcitonins are partially degraded at the site of injection, reducing their potential bioavailability. These data were obtained with an aqueous solution of SCT, the form in which it is normally administered, and suggest that absorption must be relatively efficient (Fig. 96). An attempt to improve the bioavailability of intramuscular SCT by using an aqueous gelatin vehicle instead of the normal aqueous solution was not really successful because, while the trend was in favour of the gelatin form, statistically the areas under the plasma concentration/time curves were not significantly different, although this may have been due to the relatively small number of subjects (n = 12). On the other hand, the plasma peak obtained with the gelatin vehicle was much lower than that obtained with the non-gelatin vehicle (author's unpublished observations; Fig. 97).

Metabolism

The calcitonin molecule is degraded into inactive fragments in various compartments and organs of the body, the pattern and rate of degradation appearing to differ between SCT and ECT on the one hand and HCT and PCT on the other[184,304,305]. This process has been studied *in vivo* by following the time course of the hormone's hypocalcaemic activity[306] (Figs 98–99). When incubated *in vitro* at 37°C, rat and human plasma and serum containing calcitonin showed a 50% loss of hypocalcaemic activity over a period of 6 h for SCT (and probably ECT), 2 h for HCT and 1 h for PCT (Figs 100–101). When calcitonin was incubated with extracts of liver, kidney or spleen, hypocalcaemic activity appeared to be lost most quickly from

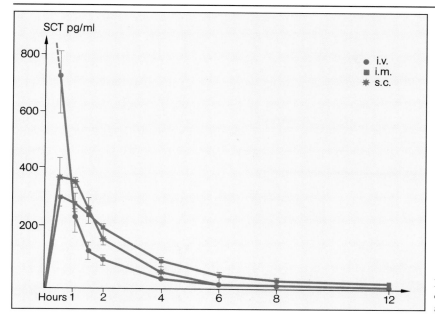

Fig. 96   Mean (± SEM) SCT plasma level curves after i.m. and s.c. compared with i.v. administration of a dose of 35 μg (n = 6)[295]

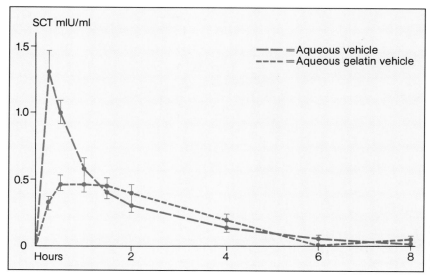

Fig. 97   Mean (± SEM) plasma concentration curves in human subjects (n = 12) after an i.m. dose of 100 IU SCT in two different vehicles

PCT, less quickly from HCT and least quickly from SCT and, probably, ECT[186,306]. SCT and (probably) ECT appear to be degraded more rapidly in kidney than in liver extracts[306] (Fig. 102). The same is probably true for HCT, but to a lesser extent.

It is thought that these differences between results obtained in different studies are partly explained by the presence in plasma, liver (rat), kidney (rat) and spleen of a factor ('calcitoninase'[186]) which is capable of degrading the calcitonins[307] but which is inactivated by

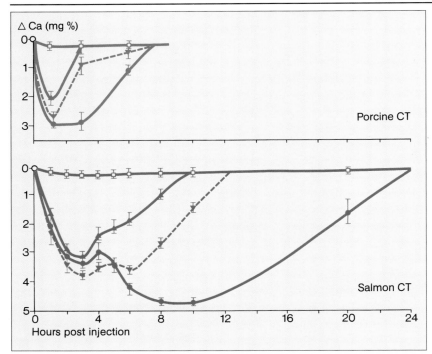

Fig. 98   Plasma calcium area response curves for porcine and salmon calcitonins in the rat. Groups of 3–5 rats were used at each dose level. The control rats were given vehicle only (—○—), while the other groups received calcitonin doses of 20 mIU/80 g bw (—▲—), 40 mIU/80 g (---▼---) or 400 mIU/80 g (—●—). The vertical bars at each point indicate SEM[306].

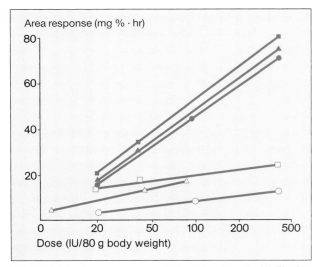

Fig. 99   Plots of the area response in the rat, as described in Fig. 98, against log calcitonin dose of hormones from several species. Note the two distinct families of curves. The mammalian hormones comprise one group (human —○—; bovine —△—; porcine —□—) and the non-mammalian hormones another (cod —●—; salmon —▲—; chicken —■—)[306].

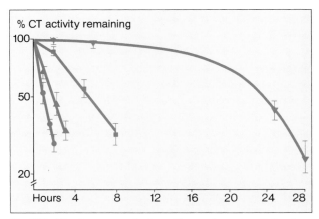

Fig. 100   The time course for loss of hypocalcaemic activity of calcitonins from several species when incubated *in vitro* in rat plasma at 37° C. All calcitonins were present at initial concentrations of 20 mIU/ml plasma. Aliquots were stored frozen until bioassayed. The vertical bars at each point indicate SEM. Chicken calcitonin activity (—▼—) was most persistent, followed by salmon (—■—), human (—▲—) and porcine (—●—). The % CT activity remaining in the incubation mixture is plotted on a log scale for clarity[306].

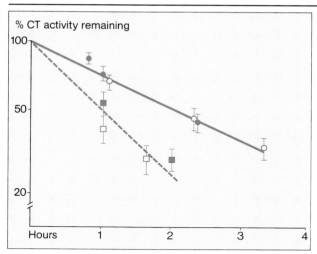

Fig. 101   A comparison of the relative stabilities of porcine and human calcitonin activity when incubated *in vitro* in rat and human plasma at 37° C. Initial calcitonin concentrations were 20 mIU/ml. Samples were stored frozen until bioassayed. The hypocalcaemic activity of human CT is lost at about the same rate in both human (—●—) and rat (—○—) plasma. Porcine hormone activity is also lost at the same rate in both human (—■—) and rat (—□—) plasma[306].

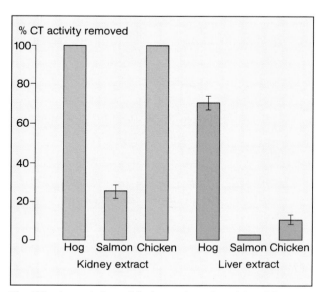

Fig. 102   A comparison of the hypocalcaemic activity of three calcitonins after incubation for 1 h at 37°C in rat liver and kidney extracts. Initial calcitonin concentrations were 20 mIU/ml and the tissue extracts contained 2.5 mg protein/ml of 0.1 M Tris-HCI buffer (pH 7.4)[306].

heat (60°C for 1 h)[306,307] (Figs 103–104). This thermolabile factor, which does not seem to be of any great importance in plasma *in vivo*, is thought to affect the properties of plasma samples not stored at low temperature[308]. *In vivo* the hypocalcaemic activity of SCT – and probably ECT, but not PCT – was more persistent in nephrectomized than in intact rats[306] (Fig. 105).

The evidence thus indicates that the primary sites of calcitonin degradation are extravascular[305-307], mainly hepatic in the case of PCT and renal in the case of HCT and SCT. Impairment of renal function consequently results in impaired elimination of these two types of hormone. SCT– and ECT, too, probably – is much more resistant to inactivation by plasma than is HCT or PCT. The possibility cannot be excluded that some degradation may occur in other organs, al-

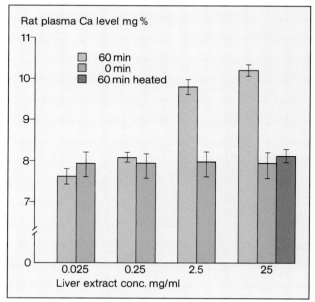

Fig. 103   The effect of rat liver extract on the hypocalcaemic activity of porcine calcitonin. Liver protein solution in 0.1 M Tris-HCI (pH 7.4) was prepared and porcine calcitonin added to a final concentration of 20 mIU/ml. Samples were removed from the incubation mixtures at zero time and after 1 h and stored frozen until bioassayed in rats. One sample of liver extract was heated to 60°C for 1 h prior to incubation with calcitonin[306].

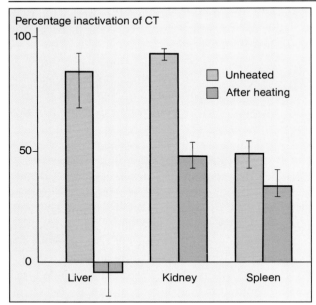

Fig. 104   Effect of heat on hepatic, renal and splenic "calcitoninases"[306,307]

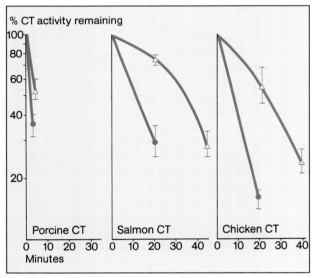

Fig. 105   The disappearance *in vivo* of hypocalcaemic activity from the plasma of anaesthetized intact (—●—) and nephrectomized (—△—) rats following a single i.v. dose of porcine, salmon or chicken calcitonin. Each rat was given 500 mIU/100 g bw of the appropriate hormone. Blood samples were collected by cardiac puncture and the plasma removed and stored frozen until bioassayed. Rats were nephrectomized 2 h prior to the experiment[306].

though there is at present no firm evidence for this. This process has also been shown in many animal species[309].

Finally, it is important for therapeutic purposes to know that synthetic salmon and human calcitonins are almost completely stable in isotonic sodium chloride and glucose infusion solutions for a period of at least 6 h (author's unpublished observations).

Summary: The biopharmaceutics of calcitonin

The main findings on the pharmacokinetics, bioavailability and metabolism of the four principal calcitonins may be summarized as follows (Fig.106):

1. *Only parenteral routes of administration (intravenous, intramuscular, subcutaneous) are at present feasible, although very recently, intranasal formulation has been made available.*

2. *SCT, the type of calcitonin most extensively studied in clinical use, has the following pharmacokinetic characteristics in doses of approximately 20 μg:*

   It is efficiently absorbed after intramuscular and subcutaneous injection, the half-life of absorption is 20–25 min and the peak plasma concentration is attained in less than 1 h.

   The half-life of elimination from plasma is 70–90 min and the metabolic clearance rate around 200 ml/min.

   The apparent distribution volume after subcutaneous injection is about 20 litres.

   Plasma protein binding is 30–40%.

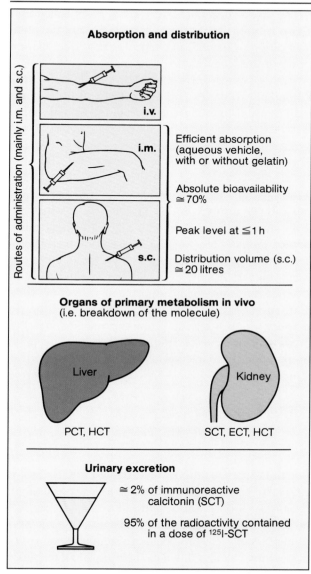

Absorption and distribution

Routes of administration (mainly i.m. and s.c.)

i.v.

i.m.

Efficient absorption
(aqueous vehicle,
with or without gelatin)

Absolute bioavailability
≅ 70%

s.c.

Peak level at ≦1 h

Distribution volume (s.c.)
≅ 20 litres

Organs of primary metabolism in vivo
(i.e. breakdown of the molecule)

Liver

Kidney

PCT, HCT

SCT, ECT, HCT

Urinary excretion

≅ 2% of immunoreactive
calcitonin (SCT)

95% of the radioactivity contained
in a dose of ^{125}I-SCT

Fig. 106   Routes of administration, absorption, distribution, metabolism and excretion

The absolute bioavailability after both subcutaneous and intramuscular injection is in the region of 70%.

The addition of gelatin to the vehicle does not significantly alter bioavailability but does reduce the peak plasma concentration.

Only about 2% of the dose is excreted as unchanged hormone (immunoreactive calcitonin) in the urine.

3. *The principal sites at which the calcitonins are degraded are extravascular:*

The liver in the case of PCT;

The kidney in the case of HCT and, especially, SCT and ECT.

4. *SCT and ECT are most resistant to catabolism and are eliminated more slowly than PCT and HCT.*

5. *Patients with renal insufficiency have a 3–5 times lower metabolic clearance rate – at least in the case of SCT – than patients with normal kidney function, necessitating individual adjustment of the dosage, especially in the presence of severe hypercalcaemia.*

6. *Differences in pharmacokinetic and metabolic characteristics do not fully explain the fact that the hypocalcaemic potency of SCT is about 30 times greater than that of HCT and PCT. Indeed, the metabolic clearance rate of SCT is only two to three times lower than that of HCT and PCT, and its greater potency is thus most likely to be the result of a higher intrinsic activity at its receptors.*

7. *Pharmacokinetic data are usually derived from the concentration of immunoreactive calcitonin and should always be interpreted in the light of the assay method used. Values obtained by radioimmunoassay tend to be low, while values based on*

radioactivity measurements (labelled CT) tend to be high. Adjustments to dosage, particularly in severe hypercalcaemia, therefore, should preferably be based on the monitoring of a clinical variable.

8. *No correlation has yet been found between plasma levels and efficacy. The peak plasma level is attained soon after administration, whereas the maximum hypocalcaemic effect occurs later and lasts longer, sometimes persisting even after the blood level has fallen below the threshold of detection (depending on the assay method used).*

9. *In man, the half-life of endogenous calcitonin is about 10 min, and the kidney is the principal site of metabolic clearance.*

Comparative summary of the principal properties of the calcitonins

Comparison of the properties of the principal calcitonins is not easy for the following reasons:

– The purity of calcitonin preparations can vary depending on the type (natural or synthetic) and manufacturing process.

– Quantities of SCT, PCT and ECT are expressed in units of hypocalcaemic activity but HCT is expressed in weight of substance, although an international unit of HCT has been defined.

– The units of SCT and PCT, which are defined in terms of their respective international reference preparations, are based solely on their hypocalcaemic effect at a given time (generally 1 h – Kumar's test) in a given species (rat). The kinetics of their hypocalcaemic effect, and therefore the area under the concentration/time curve, however, are not necessarily the same.

– The hypocalcaemic effect of a given calcitonin preparation may be, but is not necessarily, the same in different species; in particular, the effect in man may differ from that in other species.

The main properties and characteristics of the four calcitonins in clinical use are compared in Table 54.

Hypocalcaemic activity (potency)

With regard to 'intrinsic potency' the four calcitonins fall into two distinct groups, SCT and ECT, which are both derived from fish, being 20–40 times more potent (weight for weight) in Kumar's test (hypocalcaemic effect at 1 h in rats) than the mammalian calcitonins HCT and PCT. Similar plasma calcium kinetics have also been reported in dog (Fig. 107), and investigations in young rabbits[303] indicate potency ratios (derived from the areas under the blood calcium curves; Fig. 108) for the injectable preparations used in clinical practice as follows:

> HCT/SCT – 70%
> ECT/SCT – 81%
> HCT/ECT – 85%

For the three calcitonins investigated (PCT was not studied), these results confirm the following descending order of duration of hypocalcaemic effect: SCT ≥ECT > HCT. In patients with Paget's disease, SCT (and probably ECT) appears to be about ten times more active than HCT, which is in turn about ten times more potent than PCT[107]. In healthy volunteers 50 IU SCT appear to be equipotent with 75–90 IU HCT[101].

Stability in biological fluids (see Metabolism above)

SCT and ECT are much more stable than HCT and PCT in both plasma and serum and in extracts of liver, kidney and spleen[306].

Chemical stability
- In solution   SCT = ECT   > HCT   = PCT
- Oxidation sensitivity   SCT = ECT   < HCT   > PCT

Biological activity (potency)   (based on hypocalcaemic activity in rat; Kumar's test)[†]

- Units/mg   4000–6000 (SCT, ECT) 100–200 (HCT, PCT)
- 50% loss of activity in   6 h (SCT, ECT[‡])   2 h (HCT)   1 h (PCT)
- In nephrectomized rats   SCT = ECT   ≥ HCT   > PCT
- Intrinsic activity
  at receptor level   SCT = ECT   > HCT   > PCT
- Duration of hypocalcaemic
  activity   SCT = ECT   > HCT   > PCT

Affinity for receptors   SCT = ECT   > HCT   > PCT

Metabolism (degradation)
- Plasma/serum   SCT = ECT[‡] < HCT   < PCT
  (rat/man at 37°C)
- Liver (extract)   SCT = ECT[‡] < HCT   < PCT
- Kidney (extract)   SCT = ECT[‡] < HCT   < PCT

Elimination
- MCR in normal subjects   SCT = ECT   < HCT (2×) < PCT (3×)
- MCR in renal insufficiency   3–5 times lower for SCT than in normals
- Main organ involved   Kidney (SCT, ECT, HCT)   Liver (PCT)

Analgesic effect   SCT = ECT   > HCT   < PCT

Anti-inflammatory effect   SCT = ECT   > HCT   = PCT

cAMP production
- Plasma/kidney   SCT = ECT   > HCT   ≥ PCT

Diuretic effect   SCT = ECT   > HCT   = PCT

Electrolytes
- Blood/urine levels   SCT = ECT   > HCT   > PCT

Alk. phos. inhibition   SCT = ECT   > HCT   = PCT

Side effects   SCT = ECT   < HCT   = PCT
- Depend also on the percentage of impurities present in a given dose e.g. 5%
  of 100 IU = 1 μg for SCT compared with 25 μg for HCT
                               SCT less 'VIP-stimulant' than HCT?

Antibody production
- In theory   SCT = ECT[‡] > HCT   < PCT
- In practice   Not much data available for HCT
                               (for synthetic HCT, it depends on
                               whether it is identical to the natural substance
                               and on the antigenic potential of any
                               impurities

Table 54   Comparison of the principal properties of the calcitonins

† European Pharmacopoeia method
‡ Probably
= means equivalent or approximately equivalent

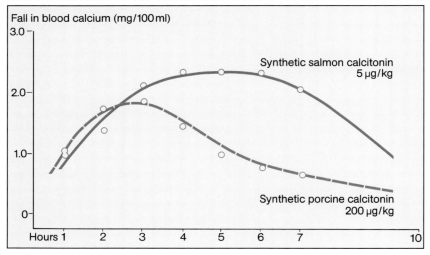

Fig. 107   Duration of effect of synthetic salmon and porcine calcitonins following i.m. injection in dogs (W.E.H. Doepfner, personal communication)

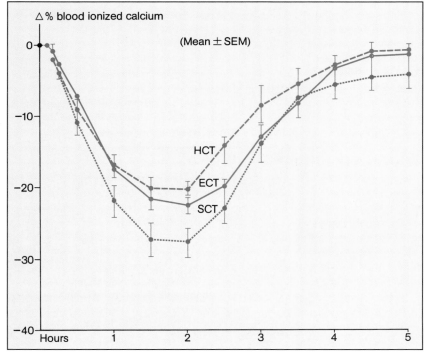

Fig. 108   Hypocalcaemic activity of three calcitonins after an i.v. dose of 1 IU/kg in young rabbits (n = 10)[303]

Interaction with receptors

SCT and ECT have greater affinity for and intrinsic activity at their receptors than do HCT and PCT.

Metabolism and elimination

The metabolic clearance rate in man is three to five times lower for SCT than for HCT[291] and PCT. On the other hand, HCT is like SCT and ECT in being de-graded mainly in the kidney, whereas PCT is broken down mainly in the liver. In nephrectomized rats, the hypocalcaemic activity of SCT is sustained longer than that of the other calcitonins; plasma levels of SCT are also sustained longer in patients with renal insufficiency.

Effects on plasma cyclic AMP

In healthy volunteers SCT induces a significant rise in plasma cyclic AMP, with a peak at 60 min[310] and a return to the baseline value at 240 min (Figs 109–110). This effect appears to be independent of the route of administration. A similar effect has been observed with HCT, with a peak at 10 min and a return to baseline at 60 min, but only after intravenous administration. SCT 50 IU had the same effect on blood levels of cyclic AMP as HCT 75–90 IU, but a more pronounced effect than ECT (S.M. Chierichetti, personal communication). Overall, therefore, SCT seems to have the greatest effect on cyclic AMP levels, although changes in plasma cyclic AMP are by no means a specific response to the activity of calcitonins. PTH, for example, may also be involved.

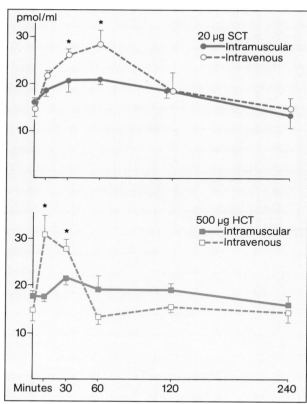

Fig. 109   Plasma concentrations of cyclic AMP in 4 healthy volunteers after administration of 20 μg SCT and of 500 μg HCT by the intramuscular and intravenous routes. Latin square experimental design (4 × 4)[310]

(*indicates p < 0.05 vs baseline.)

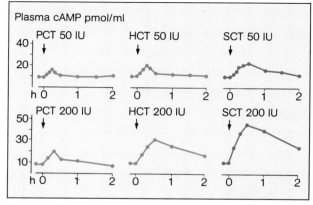

Fig. 110   Effect of intravenous infusions of biologically identical doses of PCT, HCT and SCT on plasma cAMP of two normal subjects on different days[310]

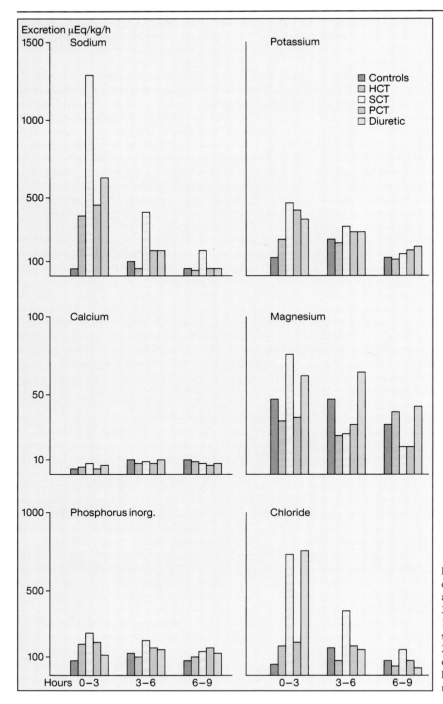

Fig. 111   Urinary excretion of various electrolytes in rats over a period of nine hours after a single subcutaneous injection of 3 mg/kg HCT or PCT, 0.02 mg/kg SCT or 1 mg/kg hydrochlorothiazide administered by stomach tube. The rats were intubated with 2 ml of 5% glucose solution at the start of the experiment. Urine was collected in three-hourly fractions. Each column represents the mean of 10 animals[71].

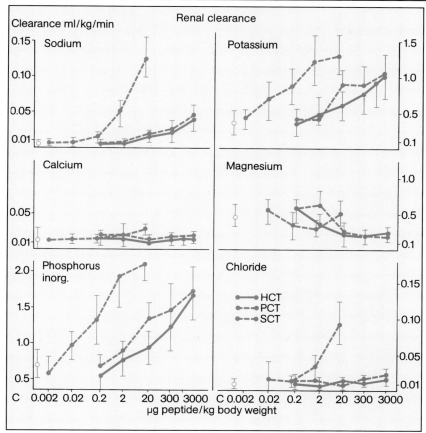

Fig. 112 Dose/response relationship for renal clearance (ml/kg/min) of various electrolytes over a period of 3 h after a single subcutaneous injection of HCT, PCT and SCT. At the start of the experiment the rats were intubated with 2 ml of 5% glucose solution. Urine was collected over 3 h. Each value represents the mean of 10 rats ± SEM[71].

Effect on β-endorphin levels

Although an effect on plasma β-endorphin levels has been reported[205,214], this action of calcitonin is unconfirmed and differences between the various calcitonins in this respect cannot usefully be discussed.

Renal effects

Direct renal effects in rat and in man[70,109] were greater after subcutaneous SCT than after PCT or HCT. SCT 50 ng had more potent saluretic and phosphaturic effects than HCT 15 μg[70] and the diuretic effect (urine volume) was also greater. Other experiments have pro-

duced similar results (Figs 111–112). SCT also had a much more pronounced effect than HCT on intracellular factors (greater stimulation of cyclic AMP and of Ca^{2+}-dependent and Mg^{2+}-dependent ATPase activity and greater inhibition of alkaline phosphatase) at renal level, as demonstrated by cytochemical methods. Its affinity for renal receptors was also greater[311].

Analgesic effects

In rats the antinociceptive effects of SCT, PCT and HCT were compared by measuring the latency of the response to pain stimuli in the hot-plate test[202]. With PCT the pain threshold was raised only slightly at 4 h,

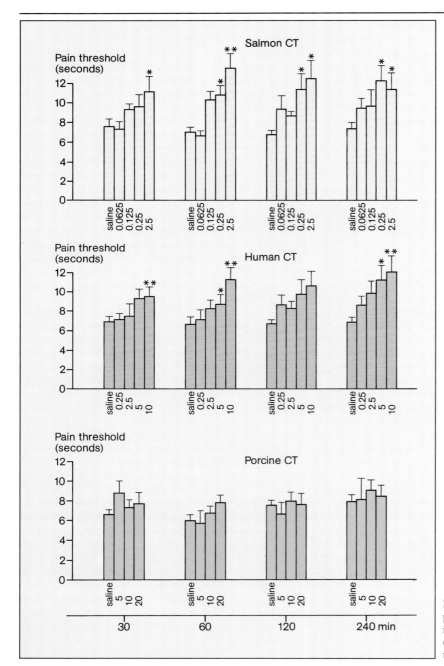

Fig. 113   Effects of salmon, human and porcine calcitonins by intracerebroventricular injection on the response to a painful stimulus (hot-plate test). Doses are in μg/rat. Dunnet t test, * p < 0.05, ** p < 0.01[202]

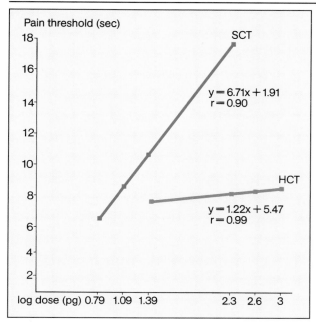

Fig. 114   Linear relation between pain threshold (hot-plate test) and log dose of SCT or HCT administered by intracerebroventricular injection to rats[202]

and only at the highest doses (10 and 20 μg/rat). HCT 5 μg raised the pain threshold by 33% at 2 h and by 37% at 4 h, and 10 μg raised the threshold by 39% at 60 min. The same effect was produced by SCT 125 ng, which is 40 times lower than the equipotent dose of HCT. After a dose of 2.5 μg SCT, the pain threshold was raised by 75% at 60 min and the analgesic effect lasted for 4 h[202] (Figs 113–114).

The analgesic activity of calcitonin seems to be bound up with its hypocalcaemic activity since SCT fragments without hypocalcaemic activity (sequences 10–16, 17–32, 24–32, 10–32) given by intracerebroventricular injection had no analgesic activity either[202].

In cancer patients 30 μg HCT and 3 μg SCT by the epidural route had substantially the same time to

onset of analgesic effect (20–30 min). However, the effect was of much shorter duration with HCT (30 min as opposed to 6-8 h with SCT)[197].

Anti-inflammatory effects

The anti-inflammatory effect of SCT is three times greater than that of HCT, which is in turn greater than that of PCT[273].

Therapeutic usefulness

Therapeutic aspects are discussed in more detail in Chapter 4 and only a brief comparison of the four calcitonins will be made at this point.

● Therapeutic effects

In *Paget's disease* SCT and HCT have comparable effects at equipotent hypocalcaemic doses, reducing bone turnover by 40–50%[312] and maintaining their effect for 3–20 months (Tables 55–56), except in a few cases where resistance develops. In *hypercalcaemia* SCT is a potent drug in short-term use and HCT appears to have a similar but less pronounced effect. Overall, however, SCT and ECT seem to be superior to the other calcitonins in this disorder (Table 57). The effects of SCT and HCT on parameters of skeletal turnover in *renal osteodystrophy* are outlined in Table 58.

No reliable comparison can be made of SCT and HCT in *Sudeck's atrophy (algoneurodystrophy), hereditary osseous dysplasia, osteogenesis imperfecta* and *primary osteoporosis* because of the small number of studies. There are also too few studies with ECT for meaningful comparison to be made with other calcitonins in the treatment of *high-bone-turnover conditions*.

Author	No. of patients	SCT dose (IU)	Average duration (months)	Mean decrease % in	
				Alkaline phosphatase	Hydroxy-proline
Bouvet[313]	46	80 qd	3	30	37
Deplante[314]	8	160 qd	3	51	29
Fornasier[315]	17	1/kg/d	12	52	–
Grunstein[316]	100	50−100 tiw−qd	9	39	46
Hadjipavlou[317]	9	50−100 qod−qd	9	40	–
Hamilton[318]	16	100 qd	5	44	40
Hosking[319]	31	100 tiw	12	57	73
Kanis[320]	7	100 tiw	6	46	–
Oreopoulos[321]	4	100−400 qd	14	69	44
Singer[322]	20	10−100 qd	12	34	35
Sturtridge[323]	28	80−100 tiw−qd	6	57	58
Trzenschik[324]	13	80 qd	6	61	31
Williams[34]	13	100 tiw	10	56	60
Woodhouse[325]	16	100 qd	12	48	70
Mean				44	44

Table 55a   Effect of synthetic salmon calcitonin on biochemical parameters of skeletal turnover in Paget's disease

Author	No. of patients	HCT dose (mg)	Average duration (months)	Mean decrease % in	
				Alkaline phosphatase	Hydroxy-proline
Burckhardt[104]	11	0.5 qd	11	33	43
Davoine[326]	4	0.5 qd	3	48	51
Epstein[327]	12	1 tiw	6	31	42
Evans[328]	41	1−2 qd	18	52	44
Gerschpacher[329]	7	0.5 qd	6	45	43
Haymovits[330]	6	0.6−1 qd	3	35	42
Murphy[331]	6	0.5 qd	20	48	53
Nuti[332]	17	0.5 qod	6	11	42
Rojanasathit[333]	5	0.5 qd	3	60	–
Woodhouse[334]	5	1 qd	9	56	66
Ziegler[335]	17	1 qd	11	50	50
Mean				42	46

Table 55b   Effect of synthetic human calcitonin on biochemical parameters of skeletal turnover in Paget's disease

Author	No. of patients	Calcitonin	Bone accretion rate (BFR)	Mean decrease % in		Skeletal mass
				Calcium pool	Osteoclast number	
Radiocalcium turnover						
Wallach[326]	11	SCT	39			
Sturtridge[323]	10	SCT	37			
Oreopoulos[321]	4	SCT	42			
Woodhouse[334]	5	HCT	38			
Caniggia[337]	8	SCT		23		
Nuti[332]	17	HCT		42		
Bone histomorphometry						
Fornasier[315]	12	SCT			40	
Singer[298]	5	SCT			65	
Williams[34]	13	SCT			44	
Haymovits[330]	2	HCT			43	
Singer[298]	4	HCT			67	
Woodhouse[334]	2	HCT			80	
Delling[338, 339]	4	HCT			47	
Partial or total body calcium						
Sturtridge[323]	16	SCT				8
Wallach[336]	10	SCT				4

Table 56   Effects of SCT and HCT on other parameters of skeletal turnover in Paget's disease

Author	No. of patients	Calcitonin	Mean decrease in total serum calcium (mmol/litre)
Nilsson[340]	12	SCT, 80−800 IU qd	0.90 ⎫
Silva[341]	23	SCT, 8−32 IU/kg/d	0.38 ⎪
Singer[322]	10	SCT, 100−1500 IU qd	0.48 ⎬ Mean 0.54
Sjöberg[342]	9	SCT, 5−40 IU/kg/d	0.80 ⎪
Wisneski[189]	24	SCT, 8−32 IU/kg/d	0.50 ⎭
Koelmeyer[343]	10	HCT, 8 mg qd	"Decreased"
Sakai[344]	3	Eel, 80−200 IU	0.11 (3 h)
Tomita[345]	8	Eel, 40−800 IU	0.13 (4−12 h)
Tomita[345]	6	Eel, 80−160 IU qd	0.54
Orimo[346]	42	[Asu^{1,7}]-eel 40−160 IU qd	0.78

Table 57   Effect of calcitonins on total serum calcium in hypercalcaemia

Author	No. of patients	Calci-tonin	Dura-tion	Mean change %			
				Serum alkaline phosphatase	Plasma hydroxy-proline	Osteo-clast number	Active resorption surface
Feletti[347]	15(9)*	SCT 1 IU kg/d	15 days each month for 6 m	↓ 35		↓ 68	↓ 57
Cundy[348]	7	SCT 10−200 IU tiw	2 m	↑ 8	↑ 16		
Farrington[349]	8 (4)*	HCT 0.5−1 mg qd	6 m	0	0	↓ 17	

Table 58   Effect of SCT and HCT on parameters of skeletal turnover in renal osteodystrophy

* Parentheses indicate number of patients biopsied

● Adverse reactions (adverse events, side effects)

Adverse reactions occur to some extent with all the calcitonins (Table 59) and may or may not be directly related to pharmacological effect. However, when considering their relative frequency and severity, it is especially important that comparisons should be made only between doses of equal biological activity and for the same route of administration. All types of the hormone have some tolerability problems when given by the intravenous route, less with intramuscular and least with subcutaneous administration. Some data suggest that HCT causes more adverse reactions than SCT and ECT after administration (especially i.v.) of doses of equal biological activity[350,351] (Tables 60–61), while SCT and ECT are about equally well tolerated (Table 62).

The whole question of adverse reactions needs more detailed study. Plasma levels of VIP were increased more by HCT than by SCT, a phenomenon that appears in many cases to be correlated with the occurrence of adverse reactions, particularly flush[350]. If, as

seems likely, at least some adverse reactions are due to the presence of byproducts or degradation products (often referred to as impurities), then HCT will inevitably tend to cause more than SCT because of the higher doses involved[352]. For example, assuming the same degree of impurity (say 5%), biologically equipotent doses of SCT and HCT would contain 1 μg and 25 μg of impurities respectively.

● Effect on antibody formation and the development of resistance

Two phenomena, antibody formation and resistance, which are sometimes associated, are dealt with in more detail in Chapter 4.

Antibodies
SCT, PCT and, above all, ECT have different aminoacid sequences from HCT, the endogenous human hormone. Normally, therefore, although antibodies may begin to be formed after long-term administration of SCT, PCT or ECT[353], this is not the case

Gastrointestinal symptoms:

- Nausea
- Vomiting
- Abdominal pain
- Diarrhoea
- Unpleasant metallic taste in the mouth

Vascular symptoms:

- Facial flushing
- Sensation of facial warmth
- Sensation of warmth affecting the hands
- Tingling in the extremities

Local symptoms:

- Erythema at the site of injection
- Pain at the site of injection

Renal symptoms:

- Increased frequency of micturition
- Polyuria

Allergic symptoms:

- Rash

Table 59   Adverse reactions occurring with calcitonins, especially after i.m. administration[351]

Type	No. of patients with side effects		Total duration of symptoms (min)		Severity: total score	
	HCT	SCT	HCT	SCT	HCT	SCT
Gastro-intestinal symptoms	3	1	105	45	21	9
Flush	8	6	300	165	52	24

Table 60a   Side effects of 100 IU HCT or SCT administered by i.v. infusion to 10 healthy volunteers (double-blind crossover study)[351]

Type	No. of patients with side effects		Total severity score	
	HCT 0.5 mg	SCT 100 IU	HCT 0.5 mg	SCT 100 IU
Gastrointestinal	6	3	8	4
Vasomotor	3	3	3	3

Table 60b   Incidence and severity of side effects in 25 patients treated with HCT 0.5 mg or SCT 100 IU[351]

Type	No. of patients with side effects			Total duration (days)			Total severity score		
	PL	HCT	SCT	PL	HCT	SCT	PL	HCT	SCT
Gastrointestinal	4	9	7	62	142	101	80	215	110
Vasomotor	2	9	7	25	156	110	32	252	149
Rash	–	–	2	–	–	23	–	–	37
Local pain	1	5	1	10	75	8	10	79	11

Table 61   Duration and severity of side effects in 36 patients (21 with metastatic osteolysis and 15 with postmenopausal osteoporosis). 12 patients received SCT (100 IU i.m./day), 12 HCT (100 IU i.m./day) and 12 placebo (PL) for 2 weeks. Gastrointestinal symptoms: diarrhoea, nausea, vomiting, abdominal pain. Vasomotor symptoms: flush, sensation of warmth, paraesthesiae[351]

Type	No. of patients with side effects		Total severity score	
	ECT 40 IU	SCT 40 IU	ECT 40 IU	SCT 40 IU
Vasomotor	1	1	1	1
Asthenia	1	1	2	1
Rash	0	1	0	1

Table 62   Incidence and severity of side effects in 12 patients treated with ECT* or SCT 40 IU[351]

* the Asu[1,7] analogue

with exogenous HCT, provided that it is identical with the natural hormone, completely pure and free from impurities which might induce antibody formation, and that it is given in doses which are unlikely to encourage such formation.

Antibodies appear after about 6 months in half to two-thirds of pagetic patients treated with porcine or salmon calcitonin[192,298,319,325,333,353-356]. Only a few cases have been reported involving HCT[299,334,357,358], but it must be remembered that HCT has not been used on anything like the scale of PCT and SCT. Those instances of antibody formation which have been reported with HCT appear to be associated with high doses. Little has yet been published on the formation of antibodies to ECT.

Resistance

Primary or secondary resistance – or non-response – which may or may not be associated with the presence of antibodies, has been reported with HCT as well as with PCT, SCT and ECT, but it is difficult to rate the different calcitonins on the basis of their relative propensity for stimulating the development of resistance.

Summary: Relative therapeutic properties

The four calcitonins used therapeutically in man may be divided into two groups: SCT and ECT in one group, HCT and PCT in the other. The biological activity of the former group is 20-40 times greater than that of the latter, the main reasons being their greater resistance to metabolic degradation – reflected in a metabolic clearance rate which is 2-3 times lower – and, above all, a high level of intrinsic activity at the target receptors.

SCT (and probably ECT) has greater analgesic potency than either HCT or PCT, but it is difficult to classify the four calcitonins with regard to side effects because of the paucity of data. Antibody production occurs with SCT, ECT and PCT, but the significance of this finding in relation to resistance is unclear, since resistance develops with both HCT and SCT even in the absence of antibodies (F. R. Singer, personal communication 1983). Overall, SCT and ECT are probably the best calcitonins for therapeutic use, followed by HCT, with PCT least suitable for long-term use in man.

References

1   Chesnut CH III et al: Calcitonin and postmenopausal osteoporosis. In: Calcitonin 1980. Proc int Symp, Milan 1980. Ed Pecile A. Excerpta Medica 1981, Int Cong Ser 540, 247–55.

2   Caniggia A et al: The rationale of calcitonin treatment in postmenopausal osteoporosis. In: Calcitonin 1980, Proc int Symp, Milan 1980. Ed Pecile A, Excerpta Medica 1981, Int Cong Ser 540, 225–36.

3   Sturtridge WC: Pharmacology and therapeutics of bone. J Dent Res 1975, 54, Spec No B, B78–85.

4   Milhaud G et al: Etude du mécanisme de l'action hypocalcémiante de la thyrocalcitonine. CR Acad Sci Paris 1965, 261, 813–6.

5   Martin TJ et al: The mode of action of thyrocalcitonin. Lancet 1966, 1, 900–2.

6   Kohler H: Wechselwirkung von Thyreocalcitonin und Parathormon. Schweiz Med Wochenschr 1968, 98, 728.

7   Munson PL: Physiology and pharmacology of thyrocalcitonin. In: Handbook of Physiology. Ed Aurbach GD. American Physiological Society, Washington 1976, 7, 443–64.

8   Chambers TJ, Dunn CJ: Pharmacological control of osteoclastic motility. Calcif Tissue Int 1983, 35, 566–70.

9   Hirsch PF, Munson PL: Thyrocalcitonin. Physiol Rev 1969, 49, 548–622.

10 Doepfner WEH: Pharmacological effects of calcitonin. Triangle (The Sandoz Journal of Medical Science) 1983, 22, 57–67.

11 Reynolds JJ, Dingle JT: A sensitive in vitro method for studying the induction and inhibition of bone resorption. Calcif Tissue Res 1970, 4, 339–49.

12 Raisz LG, Niemann I: Early effects of parathyroid hormone and thyrocalcitonin on bone in organ culture. Nature 1967, 214, 486–7.

13 Eilon G, Raisz LG: Comparison of the effects of stimulators and inhibitors of resorption on the release of lysosomal enzymes and radioactive calcium from fetal bone in organ culture. Endocrinology 1978, 103, 1969–75.

14 Aliapoulios MA, Goldhaber P, Munson PL: Thyrocalcitonin inhibition of bone resorption induced by parathyroid hormone in tissue culture. Science 1966, 151, 330–1.

15 Feldman RS, Krieger NS, Tashjian AH Jr: Effects of parathyroid hormone and calcitonin on osteoclast formation in vitro. Endocrinology 1980, 107, 1137–43.

16 Friedman J, Au WY, Raisz LG: Responses of fetal rat bone to thyrocalcitonin in tissue culture. Endocrinology 1968, 82, 149–56.

17 McLeod JF, Raisz LG: Comparison of inhibition of bone resorption and escape with calcitonin and dibutyryl 3',5' cyclic adenosine monophosphate. Endocr Res Commun 1981, 8, 49–59.

18 Raisz LG: The pharmacology of bone. Introduction. Fed Proc 1970, 29, 1176–8.

19 Reynolds JJ, Minkin C: Bone studies in vitro: Use of calcitonin as a specific inhibitor of bone resorption. In: Calcitonin 1969, Proc 2nd int Symp, London 1969. Ed Taylor S, Foster G. Heinemann 1970, 168–74.

20 MacIntyre I et al: The effect of thyrocalcitonin on blood-bone calcium equilibrium in the perfused tibia of the cat. J Physiol Lond 1967, 191, 393–405.

21 Wener JA, Gorton SJ, Raisz LG: Escape from inhibition or resorption in cultures of fetal bone treated with calcitonin and parathyroid hormone. Endocrinology 1972, 90, 752–9.

22 Tashjian AH Jr, Wright DR, Ivey JL, Pont A: Calcitonin binding sites in bone: relationships to biological response and "escape". Recent Prog Horm Res 1978, 34, 285–334.

23 Raisz LG: Bone metabolism and its hormonal regulation. Triangle (The Sandoz Journal of Medical Science) 1983, 22, 81–89.

24 Rasmussen H, Bordier P: The physiological and cellular basis of metabolic bone disease. Williams & Wilkins 1974.

25 Klein DC, Raisz LG: Role of adenosine-3', 5'-monophosphate in the hormonal regulation of bone resorption: studies with cultured fetal bone. Endocrinology 1971, 89, 818-26.

26 Luben RA, Wong GL, Cohn DV: Biochemical characterization with parathormone and calcitonin of isolated bone cells: provisional identification of osteoclasts and osteoblasts. Endocrinology 1976, 99, 526-34.

27 Murad F, Brewer HB Jr, Vaughan M: Effect of thyrocalcitonin on adenosine 3',5'-cyclic phosphate formation by rat kidney and bone. Proc Natl Acad Sci USA 1970, 65, 446-53.

28 Heersche JN et al: Calcitonin and the formation of 3',5'-AMP in bone and kidney. Endocrinology 1974, 94, 241-7.

29 Accardo G et al: Support for the clinical use of calcitonin: electron microscope study of the functional state of bone cells of rats after chronic treatment with calcitonin. Curr Ther Res Clin Exp 1982, 31, 422-33.

30 Hioco D et al: Prolonged administration of calcitonin in man: Biological, isotopic and morphological effects. In: Calcitonin 1969, Proc 2nd int Symp, London 1969. Ed Taylor S, Foster G. Heinemann 1970, 514–22.

31 Lesh JB et al: Clinical experience with porcine and salmon calcitonin. In: Endocrinology 1973, Proc 4th int Symp. Ed Taylor S. Heinemann 1974, 409–24.

32 Singer FR, Melvin KE, Mills BG: Acute effects of calcitonin on osteoclasts in man. Clin Endocrinol (Oxf) 1976, 5 Suppl, 333S–340S.

33 Epstein S, Poser J, McClintock R, Johnston CC Jr, Bryce G, Hui S: Differences in serum bone GLA protein with age and sex. Lancet 1984, 1, 307–10.

34 Williams CP, Meachim G, Taylor WH: Effect of calcitonin treatment on osteoclast counts in Paget's disease of bone. J Clin Pathol 1978, 31, 1212–7.

35 Hosking DJ et al: Bone turnover in Paget's disease – biochemical and kinetic measurement during salmon calcitonin therapy. Calcif Tissue Res 1981, 33, 471–6.

36 Holtrop ME, Raisz LG: Comparison of the effects of 1,25-dihydroxycholecalciferol, prostaglandin E2, and osteoclast-activating factor with parathyroid hormone on the ultrastructure of osteoclasts in cultured long bones of fetal rats. Calcif Tissue Int 1979, 29, 201–5.

37 Holtrop ME, Raisz LG, Simmons HA: The effects of para-

thyroid hormone, colchicine, and calcitonin on the ultrastructure and the activity of osteoclasts in organ culture. J Cell Biol 1974, 60, 346–55.

38   Kallio DM, Garant PR, Minkin C: Ultrastructural effects of calcitonin on osteoclasts in tissue culture. J Ultrastruct Res 1972, 39, 205–16.

39   Schulz A et al: Electron microscopic study on Paget-osteoclasts and the inhibitory effect of calcitonin in man. In: Molecular Endocrinology, Proc 6th int endocrinol Conf, London 1977. Ed MacIntyre I, Szelke M. Elsevier 1977, 199 ff.

40   Minaire P et al: Calcitonin treatment of acute osteoporosis resulting from paraplegia. In: Calcitonin 1984, selected short communications presented at int symp Milan 1984. Ed Doepfner WEH. Excerpta Medica 1986, Curr Clin Prac Series 42, 111–8.

41   Robinson CJ, Martin TJ, Matthews EW, MacIntyre I: Mode of action of thyrocalcitonin. J Endocrinol 1967, 39, 71–9.

42   Talmage RV, Grubb SA, Norimatsu H, Vanderwiel CJ: Evidence for an important physiological role for calcitonin. Proc Natl Acad Sci USA 1980, 77, 609–13.

43   Parfitt AM: The integration of skeletal and mineral homeostasis. In: Osteoporosis: Recent advances in pathogenesis and treatment. Ed DeLuca HF et al. University Park Press, Baltimore 1981, 115–26.

44   Toccafondi R et al: Biological effects of salmon calcitonin in osteoblast-like cells. In: Calcitonin – Chemistry, Physiology, Pharmacology and Chemical Aspects. Proc int Symp, Milan 1984. Ed Pecile A. Excerpta Medica 1985, Int Cong Ser 663, 197–204.

45   Baron R, Saffar JL: A quantitative study of the effects of prolonged calcitonin treatment on alveolar bone remodelling in the golden hamster. Calcif Tissue Res 1977, 22, 265–74.

46   Wase AW, Solewski J, Rickes E, Seidenberg J: Action of thyrocalcitonin on bone. Nature 1967, 214, 388–9.

47   Morel G et al: Immunocytochemical evidence for endogenous calcitonin and parathyroid hormone in osteoblasts from the calvaria of neonatal mice. Absence of endogenous estradiol and estradiol receptors. Cell Tissue Res 1985, 240, 89–93.

48   Komm BS et al: Bone-related cells in culture express putative estrogen receptor mRNA and 125I–17β-estradiol binding. J Bone Min Res 1987, 2, Suppl 1, abstract 237.

49   Eriksen EF: Evidence of estrogen receptors in human bone cells. J Bone Min Res 1987, 2, Suppl 1, abstract 238.

50   Eriksen EF: Multiple sex steroid receptors in cultured human osteoblast-like cells. In: Abstracts Int symp on osteoporosis, Aalborg 1987. Ed: Jensen J et al. Abstract 67.

51   Delling G et al: The effect of calcitonin on fracture healing and ectopic bone formation in the rat. In: Calcitonin 1969, Proc 2nd int Symp, London 1969. Ed Taylor S, Foster G. Heinemann 1970, 175–81.

52   Knize DM: The influence of periosteum and calcitonin on onlay bone graft survival. A roentgenographic study. Plast Reconstr Surg 1974, 53, 190–9.

53   Foster SC, Kronman JH: The effects of topical thyrocalcitonin on extraction sites in the jaws of dogs. Oral Surg Oral Med Oral Pathol 1974, 38, 866–73.

54   De Bastiani G et al: Local effects of calcitonin in bone calcification. In: Calcitonin 1980, Proc int Symp, Milan 1980. Ed Pecile A, Excerpta Medica 1981, Int Cong Ser 540, 307–13.

55   Lupulescu A: Effect of calcitonin on fibroblasts and collagen formation in rabbits: an ultrastructural and scanning electron microscopic study. J Morphol 1974, 142, 447–63.

56   Lupulescu A, Habowsky J: Effects of calcitonin on epidermal regeneration and collagen synthesis in rabbits with experimental wounds. Exp Pathol (Jena) 1978, 16, 291–302.

57   Baxter E, Fraser JR, Harris GS, Martin TJ, Melick RA: Stimulation of glycosaminoglycan synthesis by thyrocalcitonin preparations. Med J Aust 1968, 1, 216–7.

58   Kawashima K, Iwata S, Endo H: Selective activation of diaphyseal chondrocytes by parathyroid hormone, calcitonin and N_6, O_2–dibutyryl adenosine 3',5'-cyclic monophosphoric acid in proteoglycan synthesis of chick embryonic femur cultivated in vitro. Endocrinol Jpn 1980, 27, 357–61.

59   Copp DH, Kuczerpa AV: A new bioassay for calcitonin and effect of age and dietary Ca on the response. In: Calcitonin. Proc Symp on Thyrocalcitonin and the C cells, London 1967. Heinemann 1968, 18–24.

60   Denis G, Kuczerpa A: Effect of calcitonin on P uptake in epiphyseal cartilages of P-deficient rats. Can J Physiol Pharmacol 1974, 52, 355–7.

61   Talmage RV et al: The physiological significance of calcitonin. Bone min res 1983, 1, 74–143.

62   Copp DH: Modern view of the physiological role of calcitonin in vertebrates. In: The effects of calcitonin in man, Proc 1st int Wkshp, Florence 1982. Ed Gennari C, Segre G, Masson 1983, 3–12.

63   Glowacki J, Deftos LJ: The effects of calcitonin on bone formation. In: The effects of calcitonin in man, Proc 1st Int Workshop, Florence 1982, Ed Gennari C, Segre G. Masson 1983, 133–40.

64   Glowacki J, Altobelli D, Mulliken JB: Fate of mineralized and demineralized osseous implants in cranial defects. Calcif Tissue Int 1981, 33, 71–6.

65   Holtrop ME, Cox KA, Glowacki J: Cells of the mononuclear phagocytic system resorb implanted bone matrix: a histologic and ultrastructural study. Calcif Tissue Int 1982, 34, 488–94.

66   Weiss RE, Singer FR, Gorn AH, Hofer DP, Nimni ME: Calcitonin stimulates bone formation when administered prior to initiation of osteogenesis. J Clin Invest 1981, 68, 815–8.

67   Heersche JNM: Mechanism of osteoclastic bone resorption: a new hypothesis. Calcif Tissue Res 1978, 26, 81–4.

68   Heersche JNM et al: Calcitonin's actions on bone in vitro. In: Calcitonin 1980, Proc int Symp, Milan 1980. Ed Pecile A, Excerpta Medica 1981, Int Cong Ser 540, 67–78.

69   Agus ZS et al: PTH, calcitonin, cyclic nucleotides and the kidney. Ann Rev Physiol 1981, 43, 585–95.

70   Williams CC, Matthews EW, Moseley JM, MacIntyre I: The

effects of synthetic human and salmon calcitonins on electrolyte excretion in the rat. Clin Sci 1972, 42, 129–37.

71 Maier R: Pharmacology of human calcitonin. In: Human calcitonin and Paget's disease. Proc int Workshop, London 1976. Ed MacIntyre I. Huber 1977, 66–77.

72 Ardaillou R, Vuagnat P, Milhaud G, Richet G: Effects of thyrocalcitonin on the renal excretion of phosphates, calcium and hydrogen ions in man. Nephron 1967, 4, 298–314.

73 Bijvoet OL, Sluys-Veer J van der, Vries HR de, Koppen AT van: Natriuretic effect of calcitonin in man. N Engl J Med 1971, 284, 681–8.

74 Haas HG, Dambacher MA, Guncaga J, Lauffenbruger T: Renal effects of calcitonin and parathyroid extract in man. Studies in hypoparathyroidism. J Clin Invest 1971, 50, 2689–702.

75 Aldred JP, Kleszynski RR, Bastian JW: Effects of acute administration of porcine and salmon calcitonin on urine electrolyte excretion in rats. Proc Soc Exp Biol Med 1970, 134, 1175–80.

76 Aldred JP, Stubbs RK, Hermann WR, Zeedyk RA, Bastian JW: Effects of porcine calcitonin on some urine electrolytes in the rat. Acta Endocrinol (Copenh) 1970, 65, 737–50.

77 Milhaud G, Moukhtar MS: Antagonistic and synergistic actions of thyrocalcitonin and parathyroid hormone on the levels of calcium and phosphate in the rat. Nature 1966, 211, 1186–7.

78 Rasmussen H, Anast C, Arnaud C: Thyrocalcitonin, EGTA, and urinary electrolyte excretion. J Clin Invest 1967, 46, 746–52.

79 Barlet JP: Effect of porcine, salmon and human calcitonin on urinary excretion of some electrolytes in sheep. J Endocr 1972, 55, 153–61.

80 Carney S, Thompson L: Acute effect of calcitonin on rat renal electrolyte transport. Am J Physiol 1981, 240, F12–6.

81 Keeler R, Walker V, Copp DH: Natriuretic and diuretic effects of salmon calcitonin in rats. Can J Physiol Pharmacol 1970, 48, 838–41.

82 Clark JD, Kenny AD: Hog thyrocalcitonin in the dog: urinary calcium, phosphorus, magnesium and sodium responses. Endocrinology 1969, 84, 1199–205.

83 Poujeol P, Touvay C, Roinel N, de Rouffignac C: Stimulation of renal magnesium reabsorption by calcitonin in the rat. Am J Physiol 1980, 239, F524–32.

84 Puschett JB, Beck WS Jr, Jelonek A, Fernandez PC: Study of the renal tubular interactions of thyrocalcitonin, cyclic adenosine 3',5'-monophosphate, 25–hydroxycholecalciferol, and calcium ion. J Clin Invest 1974, 53, 756–67.

85 Ardaillou R, Fillastre JP, Milhaud G, Rousselet F, Delaunay F, Richet G: Renal excretion of phosphate, calcium and sodium during and after a prolonged thyrocalcitonin infusion in man. Proc Soc Exp Biol Med 1969, 131, 56–60.

86 Kawamura J, Daizyo K, Hosokawa S, Yoshida O: Acute effects of salmon calcitonin on renal electrolyte excretion in intact, thyroparathyroidectomized and sulfacetylthiazole-induced uremic rats. Nephron 1978, 21, 334–44.

87 Berndt TJ, Knox FG: Effects of parathyroid hormone and calcitonin on electrolyte excretion in the rabbit. Kidney Int 1980, 17, 473–8.

88 Koide Y, Kugai N, Yamashita K, Shimazawa E, Ogata E: A transient increase in renal clearance of phosphate in response to continuous infusion of salmon calcitonin in rats. Endocrinol Jpn 1976, 23, 295–304.

89 Quamme GA: Effect of calcitonin on calcium and magnesium transport in rat nephron. Am J Physiol 1980, 238, E573–8.

90 Sörensen OH, Hindberg I: The acute and prolonged effect of porcine calcitonin on urine electrolyte excretion in intact and parathyroidectomized rats. Acta Endocrinol (Copenh) 1972, 70, 295–307.

91 Talmage RV, Grubb SA: The influence of endogenous or exogenous calcitonin on daily urinary calcium excretion. Endocrinology 1977, 101, 1351–7.

92 Wong NLM et al: Calcitonin and renal electrolyte transport. Clin Res 1977, 25, 708A

93 Suki WN, Rouse D: Hormonal regulation of calcium transport in thick ascending limb renal tubules. Am J Physiol 1981, 241, F171–4.

94 Ardaillou R, Milhaud G, Rousselet F, Vuagnat P: Effect of thyrocalcitonin on the renal excretion of sodium and chlorine in the normal human. CR Acad Sci (D) (Paris) 1967, 264, 3037–40.

95 Langer B, Peytremann A, Rufener C, Jenny M: Comparative effects of single-dose administration of synthetic calcitonin (human and salmon) in normal subjects and patients with Paget's disease or hypercalcemia. Schweiz Med Wochenschr 1971, 101, 69–80.

96 Loreau N, Lepreux C, Ardaillou R: Calcitonin-sensitive adenylate cyclase in rat renal tubular membranes. Biochem J 1975, 150, 305–14.

97 Chabardes D, Imbert-Teboul M, Montegut M, Clique A, Morel F: Distribution of calcitonin-sensitive adenylate cyclase activity along the rabbit kidney tubule. Proc Natl Acad Sci USA 1976, 73, 3608–12

98 Kurokawa K, Nagata N, Sasaki M, Nakane K: Effects of calcitonin on the concentration of cyclic adenosine 3',5'-monophosphate in rat kidney in vivo and in vitro. Endocrinology 1974, 94, 1514–8.

99 Marx SJ et al: Calcitonin receptors of kidney and bone. Science 1972, 178, 999–1001.

100 Ardaillou R, Isaac R, Nivez MP, Kuhn JM, Cazor JL, Fillastre JP: Effect of salmon calcitonin on renal excretion of adenosine 3', 5' monophosphate in man. Horm Metab Res 1976, 8, 136–40.

101 Gennari C et al: Acute effects of salmon, human and porcine calcitonin on plasma calcium and cyclic AMP levels in man. Curr ther Res 1981, 30, 1024–32.

102 Haddad JG, Rojanasathit S: Effective human calcitonin therapy following immunological resistance to salmon calcitonin therapy in Paget's bone disease. In: Human calcitonin and Paget's disease, Proc int Workshop, London, 1976. Ed MacIntyre I. Huber, Bern 1977, 195–206.

103 Reiner M, Woodhouse NJ, Kalu DN, Foster GV, Galante L, Joplin GF, MacIntyre I: Kinetic and metabolic investigations in Paget's disease of bone, assessment of human synthetic calcitonin therapy. Helv Med Acta (Suppl) 1970, Suppl 50:136.

104 Burckhardt P et al: Treatment of Paget's disease with human calcitonin. In: Human calcitonin and Paget's disease, Proc int Workshop, London, 1976. Ed MacIntyre I. Huber 1977, 155–66.

105 Illig R, Budliger H, Kind HP, Fanconi A, Prader A: Effect of synthetic human, porcine, and salmon calcitonin on calcium and phosphorus in immobilized children. Helv Paediatr Acta 1972, 27, 225–38.

106 Paillard F, Ardaillou R, Malendin H, Fillastre JP, Prier S: Renal effects of salmon calcitonin in man. J Lab Clin Med 1972, 80, 200–16.

107 Galante L et al: The calcium lowering effect of synthetic human, porcine and salmon calcitonin in patients with Paget's disease. Clin Sci 1973, 44, 605–10.

108 Streifler C, Harell A: Proceedings: Effect of thyrocalcitonin and parathyroid hormone on adenosine triphosphatases and phosphatases of rat kidney plasma membranes. Isr J Med Sci 1975, 11, 1220–1.

109 Galante L, Horton R, Joplin GF, Woodhouse NJ, MacIntyre I: Comparison of human, porcine and salmon synthetic calcitonins in man and in the rat. Clin Sci 1971, 40, 9P–10P.

110 Pak CY, Ruskin B, Casper A: Renal effects of porcine thyrocalcitonin in the dog. Endocrinology 1970, 87, 262–70.

111 Stevenson JC, Evans IM: Pharmacology and therapeutic use of calcitonin. Drugs 1981, 21, 257–72.

112 Austin LA, Heath H 3d: Calcitonin: physiology and pathophysiology. N Engl J Med 1981, 304, 269–78.

113 Borle AB: Calcium metabolism at the cellular level. Fed Proc 1973, 32, 1944–50.

114 Borle AB: Regulation of cellular calcium metabolism and calcium transport by calcitonin. J Membr Biol 1975, 21, 125–46.

115 Borle AB: Regulation of the mitochondrial control of cellular calcium homeostasis and calcium transport by phosphate, parathyroid hormone, calcitonin, vitamin D and cyclic adenosine monophosphate. In: Calcium-regulating hormones. Ed Talmage RV, Owen M, Parsons JA. Int Cong Ser 346. Excerpta Medica, 1975, 217–28.

116 Borle AB: Control, modulation, and regulation of cell calcium. Rev Physiol Biochem Pharmacol 1981, 90, 13–153.

117 Galante L, Colston KW, MacAuley SJ, MacIntyre I: Effect of calcitonin on vitamin D metabolism. Nature 1972, 238, 271–3.

118 Lorenc R, Tanaka Y, DeLuca HF, Jones G: Lack of effect of calcitonin on the regulation of vitamin D metabolism in the rat. Endocrinology 1977, 100, 468–72.

119 Kawashima H, Torikai S, Kurokawa K: Calcitonin selectively stimulates 25–hydroxyvitamin D3–1alpha-hydroxylase in proximal straight tubule of rat kidney. Nature 1981, 291, 327–9.

120 MacIntyre I: The physiological actions of calcitonin. Triangle (The Sandoz Journal of Medical Science) 1983, 22, 69–74.

121 Larkins RG, Colston KW, Galante LS, MacAuley SJ, Evans IM, MacIntyre I: Regulation of vitamin-D metabolism without parathyroid hormone. Lancet 1973, 2, 289–91.

122 Pearse AGE et al: The neural crest origin of the C cells and their comparative cytochemistry and ultrastructure in the ultimobranchial gland. In: Calcium, parathyroid hormone and the calcitonins. Proc 4th parathyroid Conf, Chapel Hill (NC) 1971. Ed Talmage RV, Munson PL. Excerpta Medica 1972, Int Cong Ser 243, 29–40.

123 Doepfner WEH, Briner U: Calcitonin and gastric secretion. In: Calcitonin 1980, Proc int Symp, Milan 1980. Ed Pecile A, Excerpta Medica 1981, Int Cong Ser 540, 123–35.

124 Doepfner WEH, Ohnhaus EE: Gastrointestinal effects of synthetic salmon calcitonin in the cat. Biol Gastro-enterol 1972, 5, 456c.

125 Doepfner W et al: Effects of synthetic salmon calcitonin on gastric secretion and ulcer formation in conscious cats and rats. In: Effects of calcitonin and somatostatin on gastrointestinal tract and pancreas. Ed Goebell H, Hotz J. Demeter Verlag 1976, 60–70.

126 Ziegler R, Minne H, Hotz J, Goebell H: Inhibition of gastric secretion in man by oral administration of calcitonin. Digestion 1974, 11, 157–60.

127 Hotz J, Goebell H: Long-term effects of calcitonin on gastric secretion in normals, peptic ulcer and high risk patients. Z Gastroenterol (Verh) 1976, 71–7.

128 Hotz J, Goebell H, Hirche H, Minne H, Ziegler R: Inhibition of human gastric secretion by intragastrically administered calcitonin. Digestion 1980, 20, 180–9.

129 Hesch RD, Hufner M, Hasenhager B, Creutzfeldt W: Inhibition of gastric secretion by calcitonin in man. Horm Metab Res 1971, 3, 140.

130 Hotz J, Goebell H: Pharmacological actions of calcitonin on the gastrointestinal tract and their therapeutical implications. Z Gastroenterol 1981, 19, 637–45.

131 Iwatsuki K, Hashimoto K: Effects of calcitonin on the secretion of pancreatic juice induced by dopamine, secretin and pancreozymin. Clin Exp Pharmacol Physiol 1976, 3, 159–65.

132 Becker HD, Reeder DD, Scurry MT, Thompson JC: Inhibition of gastrin release and gastric secretion by calcitonin in patients with peptic ulcer. Am J Surg 1974, 127, 71–5.

133 Paul F, Neumann F, Huchzermeyer H: Response of basal and pentagastrin-stimulated gastric secretion and of serum gastrin to short- and long-term intravenous infusion of salmon calcitonin in hyperchlorhydric subjects. Scand J Gastroenterol 1978, 13, 959–67.

134 Becker HD: Calcitonin, gastrin and gastric secretion. Z Gastroenterol (Verh) 1976, 16–22.

135 Creutzfeldt W, Lankisch PG, Folsch UR: Inhibition by somatostatin of pancreatic juice and enzyme secretion and gallbladder contraction in man induced by secretin and cholecystokinin-pancreozymin administration. Dtsch Med Wochenschr 1975, 100, 1135–8.

136   Paul F: Does salmon calcitonin influence the motility of the human gastrointestinal tract? An electromanometric and endoscopic study. Z Gastroenterol (Verh) 1976, 23–7.

137   Hotz J, Minne H, Ziegler R: Relations between excretory pancreas function and calcium metabolism in man. Verh Dtsch Ges Inn Med 1972, 78, 1420–3.

138   Schmidt H, Hesch RD, Hufner M, Paschen K, Creutzfeldt W: Calcitonin-induced inhibition of exocrine pancreatic secretion in man. Dtsch Med Wochenschr 1971, 96, 1773–5.

139   Goebell H, Ammann R, Herfarth C, Horn J, Hotz J, Knoblauch M, Schmid M, Jaeger M, Akovbiantz A, Linder E, Abt K, Nüesch E, Barth E: A double-blind trial of synthetic salmon calcitonin in the treatment of acute pancreatitis. Scand J Gastroenterol 1979, 14, 881–9.

140   Goebell H, Hotz J: Inhibition of pancreatic secretion of enzymes by calcitonin. In: Effects of calcitonin and somatostatin on gastrointestinal tract and pancreas. Ed Goebell H, Hotz J. Demeter Verlag 1976, 28–31.

141   Hotz J, Minne H, Ziegler R: The influences of acute hyper- and hypocalcemia and of calcitonin on exocrine pancreatic function in man. Res Exp Med (Berl) 1973, 160, 152–65.

142   Odes HS, Barbezat GO, Clain JE, Bank S: The effect of calcitonin on secretin-stimulated pancreatic secretion in man. S Afr Med J 1978, 53, 201–3.

143   Konturek SJ et al: Effect of calcitonin on gastric and pancreatic secretion and peptic ulcer formation in cats. Dig Dis 1974, 19, 235–41.

144   Segre G et al: Non-traditional activities of calcitonin. In: The effects of calcitonin in man, Proc 1st int Wkshp, Florence 1982. Ed Gennari C, Segre G, Masson 1983, 55–64.

145   Zofkova I, Hampl R, Nedvidkova J: Effect of calcitonin on blood levels of glucose, insulin, somatotropin and cortisol. Horm Metab Res 1984, 16, 499.

146   Ziegler R, Bellwinkel S, Schmidtchen D, Minne H: Effects of hypocalcemia, hypercalcemia and calcitonin on glucose stimulated insulin secretion in man. Horm Metab Res 1972, 4, 60.

147   Minne H, Bellwinkel S, Ziegler R: The effect of calcitonin on glucose assimilation and insulin secretion in man. Acta Endocrinol (Suppl) (Kbh) 1973, 173, 162.

148   Gattereau A, Bielmann P, Durivage J, Davignon J, Larochelle P: Effect of acute and chronic administration of calcitonin on serum glucose in patients with Paget's disease of bone. J Clin Endocrinol Metab 1980, 51, 354–7.

149   Cantalamessa L, Catania A, Reschini E, Peracchi M: Inhibitory effect of calcitonin on growth hormone and insulin secretion in man. Metabolism 1978, 27, 987–92.

150   Ziegler R et al: Calcitonin and the endocrine pancreas. In: Calcitonin 1980, Proc int Symp, Milan 1980. Ed Pecile A, Excerpta Medica 1981, Int Cong Ser 540, 314–25.

151   Gray TK, Brannan P, Juan D, Morawski G, Fordtran JS: Ion transport changes during calcitonin-induced intestinal secretion in man. Gastroenterology 1976, 71, 392–8.

152   Kisloff B, Moore EW: Effects of intravenous calcitonin on water, electrolyte, and calcium movement across in vivo rabbit jejunum and ileum. Gastroenterology 1977, 72, 462–8.

153   Walling MW, Brasitus TA, Kimberg DV: Effects of calcitonin and substance P on the transport of Ca, Na and Cl across rat ileum in vitro. Gastroenterology 1977, 73, 89–94.

154   Corradino RA: Parathyroid hormone and calcitonin: no direct effect on vitamin D3–mediated intestinal calcium absorptive mechanism. Horm Metab Res 1976, 8, 485–8.

155   Cramer CF, Parkes CO, Copp DH: The effect of chicken and hog calcitonin on some parameters of Ca, P, and Mg metabolism in dogs. Can J Physiol Pharmacol 1969, 47, 181–4.

156   Cramer CF: Effect of salmon calcitonin on in vivo calcium absorption in rats. Calcif Tissue Res 1973, 13, 169–72.

157   Gray TK, Bieberdorf FA, Fordtran JS: Thyrocalcitonin and the jejunal absorption of calcium, water, and electrolytes in normal subjects. J Clin Invest 1973, 52, 3084–8.

158   Hehrmann R, Hagemann J, Montz R, Jentsch E: Differential action of parathyroid and thyroid hormones on effective intestinal absorption of calcium. In vivo studies with 47-calcium in rats. Acta Endocrinol (Copenh) 1973, 73, 489–98.

159   Loreau N, Blain F, Ardaillou R: Effects of calcitonin on phosphate, calcium and sodium transfer through the jejunum in rats. J Physiol (Paris) 1971, 63, 249–50A.

160   Bernier JJ, Rambaud JC, Cattan D, Prost A: Diarrhoea associated with medullary carcinoma of the thyroid. Gut 1969, 10, 980–5.

161   Cox TM, Fagan EA, Hillyard CJ, Allison DJ, Chadwick VS: Role of calcitonin in diarrhoea associated with medullary carcinoma of the thyroid. Gut 1979, 20, 629–33.

162   Gruson M, Cerland M, Miravet L, Paul J, Hioco D: Action of porcine calcitonin on the in vitro transfer of Ca-45 at the duodenum level of normal and parathyroidectomized rats. C R Acad Sci (D) (Paris) 1970, 270, 1014–7.

163   Milhaud G, Moukhtar MS: Thyrocalcitonin: effects on calcium kinetics in the rat. Proc Soc Exp Biol Med 1966, 123, 207–9.

164   McKercher HG, Radde IC: The effect of porcine calcitonin on intestinal calcium and phosphate fluxes in the young piglet. Can J Physiol Pharmacol 1981, 59, 71–5.

165   Heuer LJ: Inhibition of gastroenteral absorption of 45Ca by synthetic salmon calcitonin. Arzneimittelforschung 1982, 32, 32–4.

166   Swaminathan R, Ker J, Care D: Calcitonin and intestinal calcium absorption. J Endocrinol 1974, 61, 83–94.

167   Olson EB et al: The effect of calcitonin and parathyroid hormone on calcium transport of isolated intestine. In: Calcium, parathyroid hormone and the calcitonins. Proc 4th parathyroid Conf, Chapel Hill (NC) 1971. Ed Talmage RV, Munson PL. Excerpta Medica 1972, Int Cong Ser 243, 240 ff.

168   Olson EB Jr, Deluca HF, Potts JT Jr: Calcitonin inhibition of vitamin D-induced intestinal calcium absorption. Endocrinology 1972, 90, 151–7.

169   Robinson CJ et al: The effect of parathyroid hormone and thyrocalcitonin on the intestinal absorption of calcium and magnesium. In: Les tissus calcifiés. Proc 5th European

Symp, Paris 1967. Ed Milhaud G et al. Société d'édition d'enseignement supérieur 1968, 279–82

170   Goebell H, Hotz J: Calcitonin, pancreatic secretion and pancreatitis. In: Calcitonin 1980, Proc int Symp, Milan 1980. Ed Pecile A, Excerpta Medica 1981, Int Cong Ser 540, 346–51.

171   Morley JE, Levine AS, Silvis SE: Intraventricular calcitonin inhibits gastric acid secretion. Science 1981, 214, 671–3.

172   Donowitz M, Charney AN: Effect of chronic elevation of blood serotonin on intestinal transport. Gastroenterology 1976, 70, 880.

173   Hano J et al: The influence of serotonin on insulin-stimulated gastric secretion, blood glucose and serum electrolyte levels in the unanesthetized rat. Arch int Pharmacodyn 1975, 216, 28–39.

174   Nakhla AM, Latif A: A possible role for 5–hydroxytryptamine as a mediator for calcitonin actions on the gastrointestinal tract and pancreas in rats. Biochem J 1978, 176, 339–42.

175   Nakhla AM: Effect of calcitonin on acetylcholinesterase activity in the gastrointestinal tract and pancreas. J Neural Transmission 1980, 47, 313–7.

176   Goto Y et al: Antiulcer mechanisms of calcitonin: central inhibition of acid secretion through the modulation of prostaglandins. Gastroenterology 1985, 88, 1401.

177   Hotz J et al: Behandlungsmöglichkeiten des Zollinger-Ellison-Syndroms mit Kalzitonin. Therapiewoche 1978, 28, 1066–75.

178   Luisetto G et al: Treatment of acute edematous pancreatitis and pancreatic fistulas with salmon calcitonin. In: Calcitonin 1980, Proc int Symp, Milan 1980. Ed Pecile A, Excerpta Medica 1981, Int Cong Ser 540, 336–45.

179   Paul F, Ohnhaus EE, Hesch RD, Chemnitz G, Hoppe-Seyler R, Henrichs HR, Hartung H, Waldmann D, Kunze K, Barth E, Nüesch E, Abt K: Influence of salmon calcitonin on acute pancreatitis. Dtsch Med Wochenschr 1979, 104, 615–22.

180   Gennari C et al: Gli effetti collaterali di differenti calcitonine. In: The effects of calcitonin in man, Proc 1st int Wkshp, Florence 1982. Ed Gennari C, Segre G, Masson 1983, 93–101.

181   Azria M, Kiger JL: Nouvelle technique de dosage biologique de la calcitonine par utilisation du porc miniature. Thérapie 1974, 29, 753–66.

182   Fournie A et al: Test d'hypocalcémie aiguë à la calcitonine de porc et du saumon. Rev Rhum 1977, 44, 91–8.

183   Ziliotto D et al: Effetti della calcitonina sul calcio, magnesio e fosforo plasmatici in rapporto con la velocità del ricambio osseo. Minerva endocrinol 1976, 1, 159–68.

184   Chapuy MC, Meunier PJ: Comparison of the acute effect of eel and salmon calcitonins in pagetic patients. Horm Metab Res 1982, 14, 559–60.

185   Chapuy MC, Meunier PJ: Antiosteoclastic effect of calcitonin. Effect of a weak dose (letter). Nouv Presse Med 1981, 10, 2210–1.

186   Milhaud G: Calcitonine. In: Pharmacologie clinique, Vol 1.

Ed Giroud JP et al. Expansion scientifique française, Paris 1978, 859–71.

187   Stevenson JC: The structure and function of calcitonin. Invest Cell Pathol 1980, 3, 187–93.

188   Milhaud G, Job JC: Thyrocalcitonin: effect on idiopathic hypercalcemia. Science 1966, 154, 794–6.

189   Wisneski LA, Croom WP, Silva OL, Becker KL: Salmon calcitonin in hypercalcemia. Clin Pharmacol Ther 1978, 24, 219–22.

190   Rico H et al: Treatment of postmenopausal osteoporosis with calcitonin and calcium. Long-term results. In: Osteoporosis, social and clinical aspects, Proc 2nd int Conf, Athens 1985, Masson 1986, 376–80.

191   Deftos LJ, Parthemore JG, Price PA: Changes in plasma bone GLA protein during treatment of bone disease. Calcif Tissue Int 1982, 34, 121–4.

192   DeRose J et al: Response of Paget's disease to porcine and salmon calcitonins: effects of long-term treatment. Am J Med 1974, 56, 858–66.

193   Olgiati VR, Guidobono F, Luisetto G, Netti C, Bianchi C, Pecile A: Calcitonin inhibition of physiological and stimulated prolactin secretion in rats. Life Sci 1981, 29, 585–94.

194   Freed WJ, Perlow MJ, Wyatt RJ: Calcitonin: inhibitory effect on eating in rats. Science 1979, 206, 850–2.

195   Levine AS, Morley JE: Reduction of feeding in rats by calcitonin. Brain Res 1981, 222, 187–91.

196   Gibbs J, Smith GP: The neuroendocrinology of satiety. Frontiers in neuroendocrinology 1984, 8, 223–45.

197   Fiore CE et al: Alcune attivita' extrascheletriche delle calcitonine: Effetto antalgico e neuromodulatore. In: The effects of calcitonin in man, Proc 1st int Wkshp, Florence 1982. Ed Gennari C, Segre G, Masson 1983, 291–300.

198   Bijvoet OLM, Jansen AP: Thyrocalcitonin in Paget's disease (letter). Lancet 1967, 2, 471–2.

199   Welzel D: Analgesic potential of salmon calcitonin in postoperative pain. In: The effects of calcitonin in man, Proc 1st int Wkshp, Florence 1982. Ed Gennari C, Segre G, Masson 1983, 223–32.

200   Braga P, Ferri S, Santagostino A, Olgiati VR, Pecile A: Lack of opiate receptor involvement in centrally induced calcitonin analgesia. Life Sci 1978, 22, 971–7.

201   Bates RFL et al: Comparison of the analgesic effects of subcutaneous and intracerebroventricular injection of calcitonin on acetic acid-induced abdominal constrictions in the mouse. Br J Pharmacol 1981, 72/3, 575P.

202   Pecile A et al: Attivitata analgesica di calcitonine di diversa origine. In: The effects of calcitonin in man, Proc 1st int Wkshp, Florence 1982. Ed Gennari C, Segre G, Masson 1983, 205–11.

203   Yamamoto M, Kumagai F, Tachikawa S, Maeno H: Lack of effect of levallorphan on analgesia induced by intraventricular application of porcine calcitonin in mice. Eur J Pharmacol 1979, 55, 211–3.

204   Bates RFL et al: The interaction of naloxone and calcitonin

in the production of analgesia in the mouse. Br J Pharmacol 1981, 74, 279P.

205 Gennari C et al: Dolore osseo, endorfine e calcitonine. In: The effects of calcitonin in man, Proc 1st int Wkshp, Florence 1982. Ed Gennari C, Segre G, Masson 1983, 213–22.

206 Allan E: Calcitonin in the treatment of intractable pain from advanced malignancy. Pharmatherapeutica 1983, 3, 482–6.

207 Szanto J, Sandor J: Preliminary observations on the analgesic effect of salmon calcitonin in osteolytic metastases. Clin Trials J 1983, 20, 266–74.

208 Hindley AC, Hill EB, Leyland MJ, Wiles AE: A double-blind controlled trial of salmon calcitonin in pain due to malignancy. Cancer Chemother Pharmacol 1982, 9, 71–4.

209 Fiore CE, Castorina F, Malatino LS, Tamburino C: Antalgic activity of calcitonin: effectiveness of the epidural and subarachnoid routes in man. Int J Clin Pharmacol Res 1983, 3, 257–60.

210 Kleibel F et al: Acute analgesic effect of salmon calcitonin in patients with bone metastases. Neurosci Lett 1983, Suppl 14, S199.

211 Gennari C et al: Effects of calcitonin treatment on bone pain and bone turnover in Paget's disease of bone. Min Metab Res It 1981, 2, 109–13.

212 Pecile A, Ferri S, Braga PC, Olgiati VR: Effects of intracerebroventricular calcitonin in the conscious rabbit. Experientia 1975, 31, 332–3.

213 Di Silverio F et al: The use of calcitonin in the management of metastasizing prostatic carcinoma. In: The effects of calcitonin in man, Proc 1st int Wkshp, Florence 1982. Ed Gennari C, Segre G. Masson 1983, 165–74.

214 Laurian L et al: Calcitonin-induced increase in ACTH, β-endorphin and cortisol secretion. Horm metab Res 1986, 18, 268–71.

215 Tseng LF, Loh HH, Li CH: Beta-endorphin as a potent analgesic by intravenous injection. Nature 1976, 263, 239–40.

216 Loh HH, Li CH: Biologic activities of beta-endorphin and its related peptides. Ann NY Acad Sci 1977, 297, 115–30.

217 Editorial: How does acupuncture work? Br med J 1981, 283, 746–8.

218 Rapoport SI, Klee WA, Pettigrew KD, Ohno K: Entry of opioid peptides into the central nervous system. Science 1980, 207, 84–6.

219 Abdullahi SE, De Bastiani G, Nogarin L, Velo GP: Effect of calcitonin on carrageenan foot oedema. Agents Actions 1975, 5, 371–3.

220 Ceserani R, Colombo M, Olgiati VR, Pecile A: Calcitonin and prostaglandin system. Life Sci 1979, 25, 1851–5.

221 Gennari C et al: High plasma PGE2 levels in hypercalcemia of neoplastic diseases: effect of calcitonin. Min Metab Res It 1980, 1, 139 ff.

222 Sinzinger H, Peskar BA, Clopath P, Kovarik J, Burghuber O, Silberbauer K, Leithner C, Woloszczuk W: Calcitonin temporarily increases 6–oxo-prostaglandin F1 alpha-levels in man (letter). Prostaglandins 1980, 20, 611–2.

223 Bates RFL et al: Hyperalgesia induced by chronic subcutaneous injection of calcitonin. Br J Pharmacol 1981, 74, 280P.

224 Bates RFL et al: Interaction of calcium ions and salmon calcitonin in the production of analgesia in the mouse. Br J Pharmacol 1981, 73, 302–3P.

225 Satoh M, Amano H, Nakazawa T, Takagi H: Inhibition by calcium of analgesia induced by intracisternal injection of porcine calcitonin in mice. Res Commun Chem Pathol Pharmacol 1979, 26, 213–6.

226 Bates RFL et al: Antagonism of calcitonin-induced analgesia by ionophore A23187. Br J Pharmacol 1981, 74, 857P.

227 Harris RA, Loh HH, Way EL: Effects of divalent cations, cation chelators and an ionophore on morphine analgesia and tolerance. J Pharmacol Exp Ther 1975, 195, 488–98.

228 Guidobono F, Netti C, Sibilia V, Olgiati VR, Pecile A: Role of catecholamines in calcitonin-induced analgesia. Pharmacology 1985, 31, 342–8.

229 Clementi G, Prato A, Conforto G, Scapagnini U: Role of serotonin in the analgesic activity of calcitonin. Eur J Pharmacol 1984, 98, 449–51.

230 Fischer JA, Sagar SM, Martin JB: Characterization and regional distribution of calcitonin binding sites in the rat brain. Life Sci 1981, 29, 663–71.

231 Pecile et al: Calcitonins and pain perception. In: Pharmacological basis of anesthesiology: clinical pharmacology of new analgesics and anesthetics. Ed Tiengo M, Cousins MJ. Raven Press 1983, 157–65.

232 Dupuy B, Mounier J, Blanquet P: Some comments on the biological properties of calcitonin: an eventual new therapeutic utilization. Ann Endocrinol (Paris) 1977, 38, 323–6.

233 Mussini M, Agricola R, Coletti-Moia G, Fiore P, Rivolta A: A preliminary study on the use of calcitonin in clinical psychopathology. J Int Med Res 1984 12, 23–9.

234 Fiore CE et al: Neuroendocrine modulators affecting serum calcitonin levels in man. In: Calcitonin 1980, Proc int Symp, Milan 1980. Ed Pecile A, Excerpta Medica 1981, Int Cong Ser 540, 170–82.

235 Pecile A et al: Calcitonin and control of prolactin secretion. In: Calcitonin 1980, Proc int Symp, Milan 1980. Ed Pecile A, Excerpta Medica 1981, Int Cong Ser 540, 183–98.

236 Clementi G, Prato A, Bernardini R, Nicoletti F, Patti F, De Simone D, Scapagnini U: Effects of calcitonin on the brain of aged rats. Neurobiol Aging 1983, 4, 229–32.

237 Annunziato L: Regulation of the tuberoinfundibular and nigrostriatal systems. Evidence for different kinds of dopaminergic neurons in the brain. Neuroendocrinology 1979, 29, 66–76.

238 Annunziato L: Regulatory mechanisms of tuberohypophyseal dopaminergic neurons. In: Central and peripheral regulation of prolactin function. Ed MacLeod RM, Scapagnini U. Raven Press 1980, 59–68.

239 Nicoletti F, Clementi G, Patti F, Canonico PL, Di Giorgio RM, Matera M, Pennisi G, Angelucci L, Scapagnini U: Effects of calcitonin on rat extrapyramidal motor system:

behavioral and biochemical data. Brain Res 1982, 250, 381–5.

240   Twery MJ, Obie JF, Cooper CW: Ability of calcitonins to alter food and water consumption in the rat. Peptides 1982, 3, 749–55.

241   Gaggi R, Beltrandi E, Dall-Olio R, Ferri S: Relationships between hypocalcaemic and anorectic effect of calcitonin in the rat. Pharmacol Res Commun 1985, 17, 209–15.

242   Hagemann J et al: The effect of calcitonin on serum PTH levels in man. Acta endocrinol 1977, Suppl 212, 257.

243   Ross AJ, Cooper CW, Ramp WK, Wells SA Jr: Lack of direct effects of calcitonin and parathyroid hormone on in vitro secretion of one another from rat thyroparathyroid glands. Proc Soc Exp Biol Med 1980, 163, 315–21.

244   Chapuy MC, David L, Meunier PJ: Parathyroid function during treatment with salmon calcitonin. Horm Metab Res 1980, 12, 486–7.

245   Hill CS Jr, Ibanez ML, Samaan NA, Ahearn MJ, Clark RL: Medullary (solid) carcinoma of the thyroid gland: an analysis of the M. D. Anderson Hospital experience with patients with the tumor, its special features, and its histogenesis. Medicine (Baltimore) 1973, 52, 141–71.

246   Fahrenkrug J et al: Effect of calcitonin on serum gastrin concentration and component pattern in man. J clin Endocrinol Metab 1975, 41, 149–52.

247   Decker HD et al: Inhibition of gastrin release and gastric secretion by calcitonin in patients with peptic ulcer. Am J Surg 1974, 127, 71–5.

248   Guistina G et al: The effect of short term treatment with high doses of salmon synthetic calcitonin on twenty-four hours blood glucose profile, on circadian rhythm of plasma cortisol, growth hormone and prolactin, and on mixed meal induced secretion of insulin, glucagon, growth hormone and gastrin. In: Calcitonin 1980, Proc int Symp, Milan 1980. Ed Pecile A, Excerpta Medica 1981, Int Cong Ser 540, 326–35.

249   Petralito A, Lunetta M, Liuzzo A, Fiore CE, Heynen G: Effects of salmon calcitonin on blood glucose and insulin levels under basal conditions and after intravenous glucose load. J Endocrinol Invest 1979, 2, 209–11.

250   Sgambato S et al: Effect of calcitonin on glucose-stimulated secretion in normal, obese and prediabetic subjects. Il Farmaco 1978, 33, 256–62.

251   Chiba T et al: Effects of [Asu1,7]-eel calcitonin on gastric somatostatin and gastrin release. Gut 1980, 21, 94–7.

252   Calabro A et al: Calcitonin increases peripheral plasma somatostatin-like immunoreactivity levels in humans. Boll Soc Ital Biol sper 1984, 60, 1145–51.

253   Ziliotto D et al: Decrease in serum prolactin levels after acute intravenous injection of salmon calcitonin in normal subjects. Horm metab Res 1981, 13, 64–7.

254   Cavagnini F et al: Ripercussioni della terapia con calcitonina sulla funzione ipotalamo ipofisaria e pancreatica endocrina. In: Atti, Simposio internationale sulle applicazioni terapeutiche della calcitonina, Capri 1977, 143–55.

255   Leicht E, Biro G, Weinges KF: Inhibition of releasing-hormone-induced secretion of TSH and LH by calcitonin. Horm Metab Res 1974, 6, 410–4.

256   Isaac R et al: Effects of calcitonin on basal and thyrotropin-releasing hormone-stimulated prolactin secretion in man. J Clin Endocrinol 1980, 50, 1011–5.

257   Rapisarda E, Clementi G, Fiore L, Prato A, Ceravolo A, Raffaele R, Scapagnini U: Effect of calcitonin on ACTH secretion. Pharmacol Res Commun 1984, 16, 1151–9.

258   Bruni G, Dal Pra P, Segre G: Femtogram determination of ACTH by bioradioimmunoassay. Pharmacol Res Commun 1979, 11, 853–60.

259   Carman JS, Wyatt RJ: Reduction of serum-prolactin after subcutaneous salmon calcitonin (letter). Lancet 1977, 1, 1267–8.

260   Stevenson JC, Evans IM, Colston KW, Gwee HM, Mashiter K: Serum-prolactin after subcutaneous human calcitonin (letter). Lancet 1977, 2, 711–2.

261   Fujita T et al: Calcitonin, parathyroid hormone and prolactin secretion. In: Hormonal control of calcium metabolism. Ed Cohn DV et al. Excerpta Medica 1981. Int Cong Ser 511, 287–92.

262   Olgiati VR, Netti C, Guidobono F, Pecile A: High sensitivity to calcitonin of prolactin-secretion control in lactating rats. Endocrinology 1982, 111, 641–4.

263   Chihara K et al: Role of calcitonin in regulation of growth hormone and prolactin release in rats. In: Endocrine control of bone and calcium metabolism. Proc 8th int Conf on Calcium-reg horm (The Parathyroid Conferences), Kobe (Japan) 1983. Ed Cohn DV et al. Excerpta Medica 1984, 403 (abstract).

264   Chihara K, Iwasaki J, Iwasaki Y, Minamitani N, Kaji H, Fujita T: Central nervous system effect of calcitonin: stimulation of prolactin release in rats. Brain Res 1982, 248, 331–9.

265   Iwasaki Y, Chihara K, Iwasaki J, Abe H, Fujita T: Effect of calcitonin on prolactin release in rats. Life Sci 1979, 25, 1243–8.

266   Clementi G, Nicoletti F, Patacchioli F, Prato A, Patti F, Fiore CE, Matera M, Scapagnini U: Hypoprolactinemic action of calcitonin and the tuberoinfundibular dopaminergic system. J Neurochem 1983, 40, 885–6.

267   Kordon C, Enjalbert A: Prolactin inhibiting and stimulating factors. In: Central and peripheral regulation of prolactin function. Ed MacLeod RM, Scapagnini U. Raven Press 1980, 69–77.

268   Scapagnini U et al: Prolactin effects on the brain. In: Central and peripheral regulation of prolactin function. Ed MacLeod RM, Scapagnini U. Raven Press 1980, 293–309.

269   Kaji H et al: Effect of (Asu1,7) eel calcitonin on prolactin release in normal subjects and patients with prolactinoma. Acta Endocrinol 1985, 108, 297–304.

270   Guidobono F et al: Calcitonin inhibition of prolactin release induced by serotoninergic drugs. Brit J Pharmacol 1985, 86, 687P.

271   Strettle RJ, Bates RF, Buckley GA: Evidence for a direct anti-inflammatory action of calcitonin: inhibition of his-

tamine-induced mouse pinnal oedema by porcine calcitonin. J Pharm Pharmacol 1980, 32, 192–5.

272 De Bastiani G, Nogarin L: Valutazione sperimentale dell'attivita' antinfiammatoria della calcitonina. In: The effects of calcitonin in man, Proc 1st int Wkshp, Florence 1982. Ed Gennari C, Segre G, Masson 1983, 233–8.

273 Riesterer L, Jaques R: Reduction of increased vascular permeability by calcitonin. Pharmacology 1970, 3, 53–63.

274 Gennari C: Calcitonin and bone metastases of cancer. In: Calcitonin 1980, Proc int Symp, Milan 1980. Ed Pecile A, Excerpta Medica 1981, Int Cong Ser 540, 277–87.

275 Schiano A, Acquaviva PC: Le calcitonine: Hormone vasoactive? Méditerr Méd 1976, 93, 67–9.

276 Salako LA, Smith AJ, Smith RN: The effects of porcine calcitonin on renal function in the rabbit. J Endocrinol 1971, 50, 485–91.

277 Robert L et al: Inhibition du développement de l'immuno-artériosclérose par la calcitonine. Ann biol anim Bioch Biophys 1978, 18, 195–200.

278 Robert AM, Miskulin M, Godeau G, Tixier JM, Milhaud G: Action of calcitonin on the atherosclerotic modifications of brain microvessels induced in rabbits by cholesterol feeding. Exp Mol Pathol 1982, 37, 67–73.

279 Robert L, Brechemier D, Godeau G, Labat ML, Milhaud G: Prevention of experimental immunoarteriosclerosis by calcitonin. Biochem Pharmacol 1977, 26, 2129–35.

280 Dupuy B: Antistress effects of calcitonin. Biomed Pharmacother 1983, 37, 54–7.

281 Hsu WH, Cooper CW: Hypercalcemic effect of catecholamines and its prevention by thyrocalcitonin. Calcif Tissue Res 1975, 19, 125–37.

282 Kenny AD: Effect of catecholamines on serum calcium and phosphorus levels in intact and parathyroidectomized rats. Naunyn-Schmiedebergs Arch exp Path Pharmak 1964, 248, 144–52.

283 Phillippo M, Bruce JB, Lawrence CB: The effect of adrenaline on calcitonin secretion in conscious sheep. J Endocrinol 1970, 46, 12–3.

284 Ziegler R et al: The secretion of calcitonin by the perfused ultimobranchial gland of the hen. In: Calcitonin 1969, Proc 2nd int Symp, London 1969. Ed Taylor S, Foster G. Heinemann 1970, 301–10.

285 Carman JS, Wyatt RJ: Use of calcitonin in psychotic agitation or mania. Arch Gen Psychiatry 1979, 36, 72–5.

286 Dayer JM et al: Calcitonin stimulates plasminogen activator in porcine renal tubular cells LLC-PK1. J Cell Biol 1981, 91, 195 ff.

287 Findlay DM et al: Calcitonin binding and degradation by two cultured human breast cancer cell lines (MCF 7 and T47D). Biochem J 1981, 196, 513–20.

288 Iwasaki Y, Iwasaki J, Freake HC: Growth inhibition of human breast cancer cells induced by calcitonin. Biochem Biophys Res Commun 1983, 110, 235–42.

289 Uchyama T et al: Effect of calcitonin-induced hypocalcemia on the neuromuscular and cardiovascular depressive actions of kanamycin in anesthetized and conscious rats. Arch int Pharmacodyn 1981, 249, 275–88.

290 Drack et al: Human calcitonin stimulates salivary amylase output in man. Gut 1976, 17, 620–3.

291 Nüesch E, Schmidt R: Comparative pharmacokinetics of calcitonins. In: Calcitonin 1980, Proc int Symp, Milan 1980. Ed Pecile A, Excerpta Medica 1981, Int Cong Ser 540, 352–64.

292 Ardaillou R et al: Metabolic clearance rate of radioiodinated human calcitonin in man. J Clin Invest 1970, 49, 2345–52.

293 Milhaud G, Szlamka I: Specific calcitonin-binding proteins in man. Experientia 1971, 27, 1335–6.

294 Saito Y, Yasuhara M, Okumura K, Hori R: Identification of the major binding protein of salmon calcitonin in the rat. Biochem Pharmacol 1985, 34, 3543–6.

295 Beveridge T, Niederer W, Nüesch E, Petrin A: Pharmacokinetic study with synthetic salmon calcitonin (Sandoz). Z Gastroenterol (Verh) 1976 (10), 12–5.

296 Huwyler R, Born W, Ohnhaus EE, Fischer JA: Plasma kinetics and urinary excretion of exogenous human and salmon calcitonin in man. Am J Physiol 1979, 236, E15–9.

297 Chapuy MC et al: Comparison of the acute effects of human and salmon calcitonins in Pagetic patients: relation with plasma calcitonin levels. Metab Bone Dis related Res 1980, 2, 93–7.

298 Singer FR et al: Studies of the treatment and aetiology of Paget's disease of bone. In: Human calcitonin and Paget's disease, Proc int Workshop, London 1976. Ed MacIntyre I. Huber 1977, 93–110.

299 Ardaillou R, Paillard F, Sraer J, Vallee G: Compared kinetics of salmon and human radioiodinated calcitonins in man. Horm Metab Res 1973, 5, 232–3.

300 Bijvoet OLM et al: Influence of calcitonin on renal excretion of sodium and calcium. In: Calcium, parathyroid hormone and the calcitonins. Proc 4th parathyroid Conf, Chapel Hill (NC) 1971. Ed Talmage RV, Munson PL. Excerpta Medica 1972, Int Cong Ser 243, 284–98.

301 Milhaud G et al: Studies in thyrocalcitonin. In: Calcitonin 1969, Proc 2nd int Symp, London 1969. Ed Taylor S, Foster G. Heinemann 1970, 182–93.

302 Riggs BL, Arnaud CD, Goldsmith RS, Taylor WF, McCall JT, Sessler AD: Plasma kinetics and acute effects of pharmacologic doses of porcine calcitonin in man. J Clin Endocrinol Metab 1971, 33, 115–27.

303 Azria M et al: Comparison of the hypocalcemic activities of three calcitonins using a new biological method. In: Calcitonin 1984. Proc int Symp, Milan 1984. Ed Doepfner WEH. Excerpta Medica 1986, Current Clinical Practice Series 42, 104–10.

304 De Luise M et al: Inactivation and degradation of porcine calcitonin by rat liver and relative stability of salmon calcitonin. J Endocrinol 1970, 48, 181–8.

305 Habener JF et al: Metabolism of salmon and porcine calcitonin: an explanation for the increased potency of salmon calcitonin. In: Calcium, parathyroid hormone and the calcitonins. Proc 4th parathyroid Conf, Chapel Hill (NC) 1971. Ed

Talmage RV, Munson PL. Excerpta Medica 1972, Int Cong Ser 243, 152–6.

306 Newsome FE et al: A study of the stability of calcitonin biological activity. Endocrinology 1973, 92, 1102–6.

307 Milhaud G, Hankiss J: Inactivation of thyrocalcitonin by various organs. CR Acad Sci (D) (Paris) 1969, 268, 124–7.

308 Baylin SB, Bailey AL, Hsu TH, Foster GV: Degradation of human calcitonin in human plasmas. Metabolism 1977, 26, 1345–54.

309 Foster GV et al: Metabolic fate of human calcitonin in the dog. In: Endocrinology 1971, Proc 3rd int Symp. Ed Taylor S. Heinemann 1972, 71–8.

310 Chierichetti SM et al: Farmacocinetica e farmacodinamica di differenti calcitonine. In: The effects of calcitonin in man, Proc 1st int Wkshp, Florence 1982. Ed Gennari C, Segre G, Masson 1983, 15–24.

311 Salmon DM, Azria M, Zanelli JM: Quantitative cytochemical responses to exogenously administered calcitonins in rat kidney and bone cells. Mol Cell Endocrinol 1983, 33, 293–304.

312 Wallach S: Comparative effects of salmon, human and eel calcitonins on skeletal turnover in human disease. In: The effects of calcitonin in man, Proc 1st int Wkshp, Florence 1982. Ed Gennari C, Segre G, Masson 1983, 141–51.

313 Bouvet JP: Treatment of Paget's disease with salmon thyrocalcitonin. Cooperative double-blind study. Nouv Presse Med 1976, 6, 1447–50.

314 Deplante JP, Daumont A, Bouvier M, Lejeune E: The treatment of Paget's disease using salmon thyrocalcitonin. 8 cases. Nouv Presse Med 1978, 7, 3753–6.

315 Fornasier VL, Stapleton K, Williams CC: Histologic changes in Paget's disease treated with calcitonin. Hum Pathol 1978, 9, 455–61.

316 Grunstein HS et al: Paget's disease of bone. Experiences with 100 patients treated with salmon calcitonin. Med J Aust 1981, 68, 278–80.

317 Hadjipavlou AG, Tsoukas GM, Siller TN, Danais S, Greenwood F: Combination drug therapy in treatment of Paget's disease of bone: clinical and metabolic response. J Bone Joint Surg (Am) 1977, 59, 1045–51.

318 Hamilton CR Jr: Effects of synthetic salmon calcitonin in patients with Paget's disease of bone. Am J Med 1974, 56, 315–22.

319 Hosking DJ: Calcitonin and diphosphonate in the treatment of Paget's disease of bone. Metab Bone Dis Relat Res 1981, 3, 317–26.

320 Kanis JA, Horn DB, Scott RD, Strong JA: Treatment of Paget's disease of bone with synthetic salmon calcitonin. Br Med J 1974, 3, 727–31.

321 Oreopoulos DG, Husdan H, Harrison J, Meema HE, McNeill KG, Murray TM, Ogilvie R, Rapoport A: Metabolic balance studies in patients with Paget's disease receiving salmon calcitonin over long periods. Can Med Assoc J 1977, 116, 851–5.

322 Singer FR et al: Treatment of Paget's disease of bone and hypercalcaemia with salmon calcitonin. In: Endocrinology 1973, Proc 4th int Symp, London 1973. Ed Taylor S. Heinemann 1974, 397–408.

323 Sturtridge WC, Harrison JE, Wilson DR: Long-term treatment of Paget's disease of bone with salmon calcitonin. Can Med Assoc J 1977, 117, 1031–4.

324 Trzenschik K et al: Zur Behandlung der Osteodystrophia deformans Paget mit synthetischem Salm-Kalzitonin (II. Mitteilung). Z Alternsforsch 1978, 33, 441–51.

325 Woodhouse NJ, Mohamedally SM, Saed-Nejad F, Martin TJ: Development and significance of antibodies to salmon calcitonin in patients with Paget's disease on long-term treatment. Br Med J 1977, 2, 927–9.

326 Davoine GA, Jung A, Courvoisier B: Treatment of Paget's disease of bone: diphosphonates or calcitonin? Schweiz Med Wochenschr 1981, 111, 518–24.

327 Epstein S, Owen G: Paget's disease of bone (osteitis deformans) treated with human synthetic calcitonin. S Afr Med J 1977, 51, 133–7.

328 Evans IMA: Human calcitonin in the treatment of Paget's disease: long-term trials. In: Human calcitonin and Paget's disease, Proc int Workshop, London 1976. Ed MacIntyre I. Huber 1977, 111–23.

329 Gerspacher H, Schindler H: Treatment of Paget's disease with synthetic human calcitonin. Wien Klin Wochenschr (Suppl) 1980, 92, 535–8.

330 Haymovits A et al: Studies on short-term therapy with human calcitonin and ultrastructural, immunological and genetic observations on the aetiology of Paget's disease. In: Human calcitonin and Paget's disease, Proc int Workshop, London, 1976. Ed MacIntyre I. Huber 1977, 138–54.

331 Murphy WA, Whyte MP, Haddad JG Jr: Paget bone disease: radiologic documentation of healing with human calcitonin therapy. Radiology 1980, 136, 1–4.

332 Nuti R, Vattimo A: Synthetic human calcitonin in Paget's disease of bone and osteoporosis. Dtsch Med Wochenschr 1981, 106, 149–52.

333 Rojanasathit S, Rosenberg E, Haddad JG Jr: Paget's bone disease: response to human calcitonin in patients resistant to salmon calcitonin. Lancet 1974, 2, 1412–5.

334 Woodhouse NJ, Bordier P, Fisher M, Joplin GF, Reiner M, Kalu DN, Foster GV, MacIntyre I: Human calcitonin in the treatment of Paget's bone disease. Lancet 1971, 1, 1139–43.

335 Ziegler R et al: Therapeutic studies with human calcitonin. In: Human calcitonin and Paget's disease, Proc int Workshop, London 1976. Ed MacIntyre I. Huber 1977, 167–78.

336 Wallach S, Avramides A, Flores A, Bellavia J, Cohn S: Skeletal turnover and total body elemental composition during extended calcitonin treatment of Paget's disease. Metabolism 1975, 24, 745–53.

337 Caniggia A, Gennari C, Vattimo A, Nardi P, Francini G: Therapeutic effect of synthetic calcitonin from salmon in Paget's disease and in osteoporosis. Clin Ter 1977, 82, 213–26.

338 Delling D et al: Histomorphometric studies on the influence of long-term calcitonin therapy on osteodystrophia deformans Paget. Acta Med Austriaca 1977, 4, 172–9.

339 Delling GR et al: Changes of bone remodelling surfaces and bone structure in Paget's disease following long-term treatment with calcitonin. Calcif Tiss Res 1977, 22/Suppl, 359–61.

340 Nilsson O, Almqvist S, Karlberg BE: Salmon calcitonin in the acute treatment of moderate and severe hypercalcemia in man. Acta Med Scand 1978, 204, 249–52.

341 Silva OL, Becker KL: Salmon calcitonin in the treatment of hypercalcemia. Arch Intern Med 1973, 132, 337–9.

342 Sjöberg HE, Hjern B: Acute treatment with calcitonin in primary hyperparathyroidism and severe hypercalcaemia of other origin. Acta Chir Scand 1975, 141, 90–5.

343 Koelmeyer TD, Stephens EJ: Synthetic human calcitonin in the treatment of hypercalcaemia of metastatic breast cancer: preliminary report. NZ Med J 1978, 87, 434–5.

344 Sakai T et al: Effects of synthetic eel calcitonin on hypercalcemia, with special reference to its acute effect. Clin Endocrinol Tokyo 1979, 27, 71–6.

345 Tomita A et al: Therapeutic effect of synthetic eel calcitonin on patients with hypercalcemia. Clin Endocrinol Tokyo 1979, 27, 897–902.

346 Orimo H: Clinical application of calcium-regulating hormone. Clin Endocrinol Tokyo 1979, 28, 269–74.

347 Feletti C, Bonomini V: Effect of calcitonin on bone lesions in chronic dialysis patients. Nephron 1979, 24, 85–8.

348 Cundy T et al: Responses to salmon calcitonin in chronic renal failure: relation to histological and biochemical indices of bone turnover. Eur J Clin Invest 1981, 11, 177–84.

349 Farrington K, Varghese Z, Moorhead JF: Human calcitonin in the treatment of renal osteodystrophy. J Lab Clin Med 1980, 96, 299–306.

350 Gennari C, Passeri M, Chierichetti SM, Piolini M: Side-effects of synthetic salmon and human calcitonin (letter). Lancet 1983, 1, 594–5.

351 Eisinger J, Ouaniche J: Side effects of different calcitonins. In: Abstracts for "Calcitonin in osteoporosis", Int Symp, Naxos 1986. Sandoz SpA Milan, 54–5.

352 Bangham DR, Zanelli JM: Side effects of calcitonins (letter). Lancet 1983, 1, 926–7.

353 Hosking DJ: Clinical significance of calcitonin antibodies. In: Calcitonin 1980, Proc int Symp, Milan 1980. Ed Pecile A, Excerpta Medica 1981, Int Cong Ser 540, 365–77.

354 Haddad JG Jr, Caldwell JG: Calcitonin resistance: clinical and immunologic studies in subjects with Paget's disease of bone treated with porcine and salmon calcitonins. J Clin Invest 1972, 51, 3133–41.

355 Plehwe WE, Hudson J, Clifton-Bligh P, Posen S: Porcine calcitonin in the treatment of Paget's disease of bone: experience with 32 patients. Med J Aust 1977, 1, 577–81.

356 Singer FR, Aldred JP, Neer RM, Krane SM, Potts JT Jr, Bloch KJ: An evaluation of antibodies and clinical resistance to salmon calcitonin. J Clin Invest 1972, 51, 2331–8.

357 Evans IMA et al: Paget's disease: results of long-term treatment with synthetic human calcitonin. In: Molecular endocrinology. Ed MacIntyre I, Szelke M. Elsevier 1977, 235–42.

358 Greenberg PB et al: Treatment of Paget's disease of bone with synthetic human calcitonin: biochemical and roentgenologic changes. Am J Med 1974, 56, 867–70.

359 Laurian L, Oberman Z, Hoerer E, Graf E: Antiserotonergic inhibition of calcitonin-induced increase of β-endorphin, ACTH, and cortisol secretion. J Neurol Transm 1988, 73, 167–76.

Chapter 4: Calcitonin in Therapeutic Use

Natural porcine calcitonin and synthetic salmon, eel analogue and human calcitonins are the forms of the hormone currently in therapeutic use, their principal indications being bone diseases characterized by skeletal loss and bone-renewal dysfunction, although certain disorders not involving bone have also been found to respond.

Indications

The many disorders in which calcitonin is known or thought to be therapeutically useful are listed in Table 63. Established indications include hypercalcaemic states with or without bone involvement, such as bony metastases and vitamin-D intoxication[1,2], Paget's disease[3], high-bone-turnover osteoporosis[4-7], chronic pain associated with bone disease[8], and acute pancreatitis[9].

Hypercalcaemia and hypercalcaemic crisis due to excessive osteolysis secondary to cancer of the breast, lung, kidney or other organ, or to a myeloma, hyperparathyroidism, immobilization or vitamin-D intoxication respond well to calcitonin in both acute and long-term use. *Paget's disease* (osteitis deformans) also responds well, particularly where there is bone pain with neurological complications, high bone turnover (as shown by raised serum alkaline phosphatase and urinary hydroxyproline levels) or progressive bone lesions with partial or repeated fractures.

Another classic indication for calcitonin is high-bone-turnover *osteoporosis* which can occur during primary and secondary hyperparathyroidism. In the USA, in fact, calcitonin is now qualified by the FDA as a "treatment of postmenopausal osteoporosis in conjunction with adequate calcium and vitamin-D intake to prevent the progressive loss of bone mass" (Table 64). *Bone pain* associated with osteoporosis, in particular acute or chronic pain due to vertebral fractures or osteolysis due to neoplasms, is another established indication.

A relatively recent addition to calcitonin's established uses is *algoneurodystrophy* or Sudeck's atrophy, a syndrome caused by various factors such as painful post-traumatic osteoporosis, reflex dystrophy or iatrogenic neurotrophic disorders.

In addition to these established uses of calcitonin, beneficial effects have also been reported in prevention[63], in established osteoporosis[64], in immobilization[65], in osteogenesis imperfecta[10,11], in controlling bone loss during long-term administration of prednisone or structurally similar steroids[12] or heparin, in chronic renal insufficiency associated with excessive osteoclastic activity[13], and in *acute pancreatitis* as an adjuvant to primary therapy. To these must be added a number of other possible but inadequately confirmed indications, including immunoarteriosclerosis, migraine prophylaxis, rheumatoid arthritis and periodontosis. Finally – and even more speculatively – it has been suggested that some psychiatric disorders and cardiac arrhythmias might respond to treatment with calcitonin.

Resistance

Both long-term treatment with calcitonin and *in-vitro* experiments of long duration present a number of problems. One of these is resistance[14-17], three forms of which occur (Fig. 115), primary resistance (or non-response), plateau phenomenon, and secondary (or rebound) resistance (also known as the "escape phenomenon" or as late desensitization); secondary resistance includes resistance of immunological origin associated with antibody formation.

Primary resistance or primary non-response

Some patients exhibit primary resistance to calcitonin, with no – or practically no – response to normal therapeutic doses and, often, still no response even when the dose is raised to as much as 100-500 IU/day

Indication	Efficacy
I Hypercalcaemia	
– Hypercalcaemic crisis, emergency treatment	Confirmed
– Due to primary hyperparathyroidism	Confirmed
– Due to hyperparathyroidism of neoplastic origin with or without bone metastases	Confirmed
– Due to multiple myeloma or other malignant blood cell disorders	Confirmed
– Due to thyrotoxicosis	Confirmed
– Due to immobilization	Confirmed
– Due to vitamin-D intoxication	Confirmed
– Due to drug effects (thiazide diuretics)	Confirmed
– Due to milk alkali syndrome	Probable
– Due to acute renal failure	Probable
– Due to Addison's disease	Probable
II Bone disease with decalcifying lesions	
A Generalized osteoporosis (prevention/treatment)	
– Senile	Confirmed
– Endocrine	
• Postmenopausal	Confirmed
• Hypogonadic or following castration	Confirmed
• Cushing's disease	Probable
• Hyperthyroidism	Probable
• Acromegaly	Possible
– Congenital	
• Osteogenesis imperfecta	Confirmed
• Juvenile	Probable
– Associated with GI and digestive processes	
• Postgastrectomy	Possible
• Cirrhosis of the liver	Possible
• Steatorrhoea	Possible
– Associated with neurodystrophy	
• Immobilization	Confirmed
• Neuropathy	Confirmed
– Following irradiation	Probable
– Associated with diabetes	
• Diabetes mellitus	Possible
• Haemochromatosis	Possible
– Iatrogenic	
• Heparin	Confirmed
• Cortisone	Confirmed
• Thyroxine	Confirmed
B Localized post-traumatic osteoporosis (algoneurodystrophy)	Confirmed
– Periodontosis	Probable

Indication	Efficacy
C Mixed forms (with bone destruction, decalcification and remodelling)	
– Paget's disease	Confirmed
– Metastases of prostatic carcinoma	Probable
– Metastases of other carcinomas	Probable
– Renal osteodystrophy	Confirmed
D Pain associated with bone disease	
– Chronic bone pain in osteoporosis, bone lesions of neoplastic origin, and other bone lesions	Confirmed
III Diseases of the stomach and pancreas	
– Peptic ulcer	Potential
– Acute pancreatitis	Confirmed
IV Inflammatory disorders	
– Rheumatic disease (rheumatoid arthritis)	Possible
V Cardiovascular disease	
– Migraine	Possible
– Immunoarteriosclerosis	Possible
– Raynaud's disease	Possible
– Other disorders of arterial circulation	Speculative
– Calcification of large vessels	Possible
– Arrhythmias	Speculative
– Senile cardiovascular disease	Speculative
VI Other conditions	
– Burns	Speculative
– Stress	Possible
– Psychiatric disorders	Speculative

Table 63   Therapeutic indications for calcitonin either alone or in conjuction with specific causal treatment

Prevention	Treatment
Calcitonin	Anabolic steroids
Calcium	Calcitonin
Exercise	Calcium
Oestrogens (Progestagens)	Exercise
Vitamin-D metabolites and analogues	Fluoride
	Vitamin-D metabolites and analogues

Table 64   Principal therapeutic modalities in use for osteoporosis

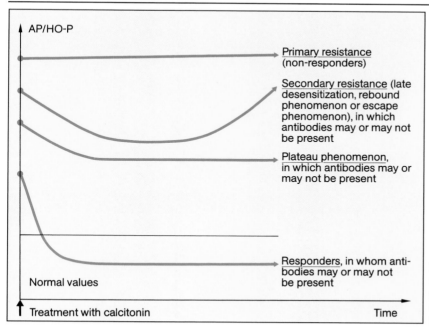

Fig. 115   Diagram illustrating the various forms of resistance to calcitonin in pagetic patients, with the associated levels of serum alkaline phosphatase and urinary hydroxyproline.

(SCT)[18] or 1–5 mg/day (HCT)[19]. Such patients, who are fortunately few, must simply be classed as non-responders.

The plateau phenomenon

In patients exhibiting the plateau phenomenon – which might be basically a biochemical phenomenon in most cases – administration of calcitonin induces a fall in serum alkaline phosphatase, but the response is inadequate and blood levels cannot be brought below a certain point whatever the type and dose of calcitonin given. Possible explanations include secondary hyperparathyroidism, the presence of antibodies to calcitonin, a reduction in the number of specific receptors, and the presence of calcitonin-resistant osteoclasts, which remain active, in addition to the normal, calcitonin-sensitive osteoclasts, which are inhibited by the hormone.

Some aspects of the plateau phenomenon are poorly understood. Why, for example, can no further re-duction in alkaline phosphatase levels be obtained in affected patients even by withdrawing and then re-instituting calcitonin, and why do some patients who exhibit the phenomenon not have antibodies to calcitonin? Above all, why do some patients exhibiting plateau phenomenon nevertheless experience continued relief of symptoms, with steady histological improvement and, radiologically, continued progress from a focal negative to a focal positive bone balance[20-23]?

Secondary resistance

In secondary resistance the initially normal inhibitory effect of calcitonin on bone resorption is lost at some stage and the patient becomes unresponsive (Figs 116–122). Sometimes there is even an exacerbation of the pathological process (rebound). Secondary resistance occurs with all the calcitonins and in about 50% of cases it is not accompanied by the formation of antibodies.

In practice, some degree of secondary resistance is

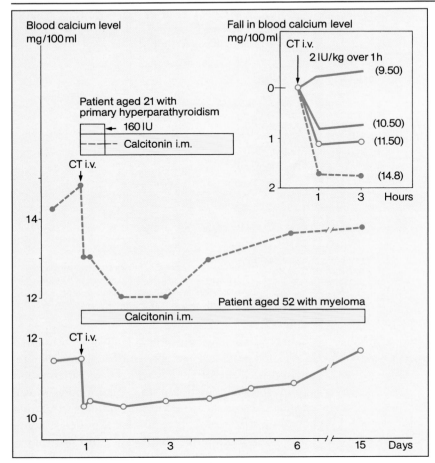

Fig. 116   Effects of calcitonin on blood calcium level, showing a noticeable decline after the first few days. The inset graph shows the hypocalcaemic response in relation to the basal blood calcium level (in brackets)[17].

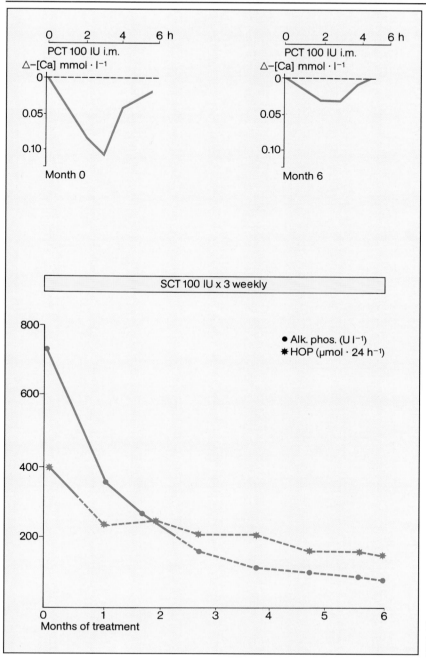

Fig. 117   Relationship between bone turnover parameters and acute hypocalcaemic response to PCT in pagetic patients[24]

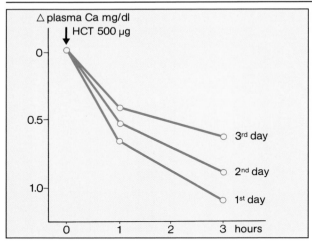

Fig. 118   Declining response of plasma calcium to (human) calcitonin over a 3-day period in patients with primary hyperparathyroidism [17]

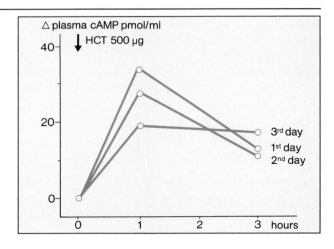

Fig. 119   Declining effect of (human) calcitonin on plasma cAMP in patients with primary hyperparathyroidism [17]

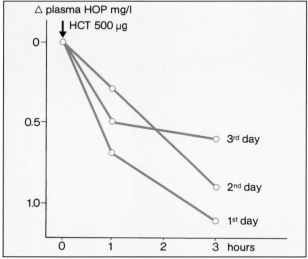

Fig. 120   Declining effect of (human) calcitonin on plasma free hydroxyproline in patients with primary hyperparathyroidism [17]

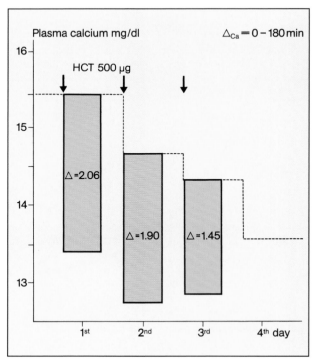

Fig. 121   Declining hypocalcaemic response to (human) calcitonin over 4 days of treatment in patients with primary hyperparathyroidism [17]

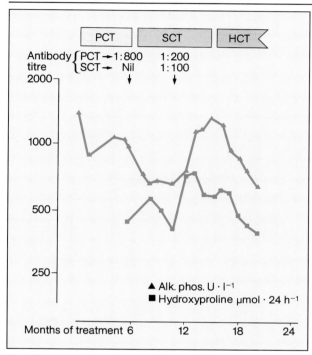

Fig. 122   Calcitonin resistance: Changes in bone turnover with different calcitonin species[24]

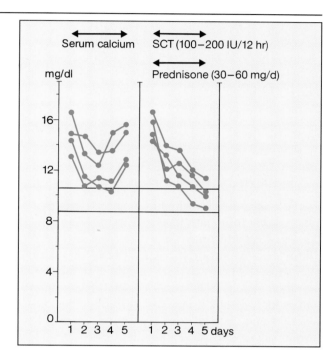

Fig. 123   Calcium levels in hypercalcaemic subjects treated with SCT alone (left) and SCT and prednisone (right)[27]

very common. Pagetic patients treated for 5 years, for example, were reported to have serum alkaline phosphatase and urinary hydroxyproline levels as much as 50% higher than the lowest levels attained, returning almost to pretreatment levels[25]. Nevertheless, the calcitonin used (HCT) in these studies remained fully active, as shown by the significant increase in both serum alkaline phosphatase and urinary hydroxyproline after treatment was stopped. Similar results have been reported for other calcitonins[26]. In patients with hypercalcaemia of neoplastic origin, e.g., salmon calcitonin 100-200 IU by intravenous infusion every 12 h induced a fall in blood calcium over the first 24 h, but from the third day onwards the hypercalcaemia returned despite continued treatment. The hypocalcaemic response was sustained, however, when calcitonin was given in combination with a glucocorticosteroid (prednisone)[27] (Fig. 123). In another study, in rats, hypocalcaemia induced by continuous infusion of calcitonin at 30 mIU/h with an Alzet pump was no longer in evidence after 24 h. In this case it was suggested that the effect of calcitonin was counteracted by PTH[28,29].

Possible mechanisms of secondary resistance
(Table 65)

One hypothesis is that the escape phenomenon is due to bone-cell differentiation and the recruitment of specific precursor cells in response to the stimulation of resorption. This is based largely on the observation that, whereas calcitonin lost its inhibitory effect on bone resorption after 24–48 h in organ cultures stimulated with PTH, calcitriol or prostaglandin E_2[30], it did not do so when cellular proliferation in the calvaria of

– Degradation (in vitro)
– Decrease in the numbers of 'functioning' receptors
– Lost or decreased affinity of CT for its specific receptors
– Blocking of receptors by inactive fragments of the calcitonin molecule or other inactive substances
– 'Uncoupling' of adenylate cyclase receptors
– Phosphate depletion
– Loss of effect of calcitonin on the differentiation of osteo-progenitor cells
– Formation of calcitonin-resistant osteoclasts
– Formation of neutralizing antibodies

Table 65   Possible mechanisms of calcitonin resistance

newborn mice was inhibited by gamma-irradiation[31]. Whole-body irradiation has been reported to inhibit the escape phenomenon in rats treated with ECT[32]. These results strongly suggest that the escape phenomenon *in vivo* involves calcitonin-induced proliferation of cells in the mononuclear phagocyte system, with a resultant increase in the number of osteoclasts and in their bone-resorptive activity. This further suggests the development of more aggressive osteoclasts that might also be more resistant to calcitonin[33] and/or the blocking of the acute inhibitory effect that calcitonin exerts under normal conditions on the process of fusion that converts certain mononuclear bone-marrow cells, known to possess calcitonin receptors, into osteoclasts[34]. This inhibitory effect can thus be 'trumped' by PTH, increasing bone resorption despite the continued presence of calcitonin – i.e. causing the escape phenomenon. Another possibility is that, after long-term exposure to calcitonin, the target cells might become insensitive, while retaining their response to PTH. This might be because there are too few receptor sites, because the receptors are occupied by inactive fragments of the calcitonin molecule, or because they have lost their affinity for calcitonin.

Whatever its cause or causes – and it is almost certainly a multifactorial phenomenon – secondary resistance may well explain why, in patients with medullary carcinoma of the thyroid, and thus permanently raised plasma calcitonin levels, bone growth and turnover continue. It may also explain why calcitonin is less effective than originally expected in the treatment of hyperparathyroidism, neoplastic hypercalcaemia and osteolytic disorders.

On the other hand, the effects seen with long-term administration of calcitonin strongly suggest that the hormone acts by inhibiting the activation of bone resorption mechanisms for long enough, at least, to have a therapeutically useful effect. That the escape phenomenon does not occur when calcitonin is given on a once-daily, intermittent basis is possibly explained by its short half-life[35]. This means that the cells that are sensitive to calcitonin are exposed only intermittently – as they are in normal endogenous secretion – reducing the likelihood of, for example, secondary hyperparathyroidism or a decrease in the number of receptors.

Another possible explanation for secondary resistance is the formation of antibodies to calcitonin (Table 66). All forms of the hormone except natural human calcitonin are potentially antigenic in human subjects exposed for any length of time because of the differences in aminoacid sequences[26,36–41]. No definite correlation has yet been detected between antibody titre and the dose of calcitonin administered[42], but it is not inconceivable that high doses, even of human calcitonin, might stimulate their production[19,43].

Antibodies have been reported to appear within 6 months of the start of treatment in 66% of pagetic patients treated with porcine calcitonin[26,36,37,40], in 50% of those treated with salmon – and presumably the Asu[1,7]-eel analogue – calcitonin[26,38–41], and in a few isolated cases with human calcitonin[19,43]. The titre tends to rise progressively the longer calcitonin is given (Fig. 124), although there are also reports of cases in which it remained constant or even declined[39]. After withdrawal of calcitonin, antibody remained detectable for 6 months in about half the patients treated[44,45]. A switch to a different type of calcitonin generally results in a fall in the titre of anti-

Author	Calcitonin	Dose (IU/week)	n	% antibodies
De Rose et al.[26]	PCT	1400	21	66
Plehwe et al.[36]	PCT	320−2240	24	43
Martin et al.[37]	PCT	560	38	28
Singer et al.[38]	SCT	70−3500	21	52
De Rose et al.[26]	SCT	150−700	28	30
Woodhouse et al.[39]	SCT	700	13	69

Table 66   Prevalence of calcitonin antibodies

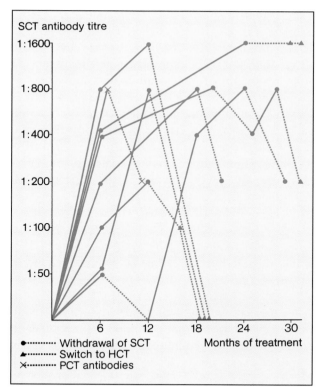

Fig. 124   Changes in antibody titres in pagetic patients following withdrawal of SCT and switch to HCT[44]

body to the type previously given[39], but changing back tends to be followed by a more rapid and higher build-up of the original titre[24,44]. Patients in whom antibodies to one calcitonin develop generally show an immunological response to another type, although there are exceptions[24,26,44,45]. Anticalcitonin antibodies are reported to belong to the 7S-immunoglobulin-G class[40].

Genuine cases of antibody-mediated resistance to calcitonin are rare and the importance of such resistance may have been grossly exaggerated[24,44]. The reported frequency varies greatly, ranging from 5-6%[26] to 40%[40], with an intermediate figure of 22% for a group of patients treated with salmon calcitonin[46]. The situation is the same as in diabetes, in fact, where antibodies to insulin are common but antibody-based resistance is rare.

Effect of resistance development

The practical relevance of calcitonin antigenicity may be assessed in three ways[24,44], although all three methods have limitations and can lead to overestimates of the extent of resistance, either for reasons inherent in the methods themselves or because of the way in which their results are interpreted[47].

● Loss of the hypocalcaemic response to antigenic calcitonin in vivo

To determine whether a patient is producing antibodies that interfere with the action of a particular calcitonin, either by blocking its receptor sites or by irreversible binding, the acute hypocalcaemic response to two different calcitonins should be compared. The response to the calcitonin to which antibodies are assumed to have formed should first be measured and this should then be compared with the response to a reference calcitonin that does not cross-react with the antibodies present. If resistance is due to antibodies, the fall in blood calcium will not occur with the first calcitonin but will be seen with the calcitonin which does not cross-react.

The only proof of the existence of antibody-mediated resistance to calcitonin in a given patient thus appears to be a resumption of the hypocalcaemic response when calcitonin of another species is substituted[47]. In practice, however, this method is less straightforward than it appears, since the loss of acute response to calcitonin resulting from the presence of antibodies must be distinguished from that due to a fall – for whatever reason – in bone turnover as, for example, in Paget's disease[48,49]. There is also an inherent variability in the acute hypocalcaemic response itself, which severely limits the reliability of this diagnostic test for antibody-mediated resistance[24].

The results indicate, however, that most patients being treated with antigenic calcitonins continue to show an acute hypocalcaemic response in line with their bone-turnover rate, and this response is largely independent of the presence of antibodies and the antibody titre. In only a very few patients does the acute hypocalcaemic response truly reflect antibody-mediated interference. In theory, it should be possible to overcome resistance of this type by giving a larger dose of calcitonin, some of which would be consumed in combining with antibodies, the remainder being free to exert its hormonal effect. It also seems

likely that in most cases binding of calcitonin to antibodies is reversible *in vivo*[50], as is the case with insulin[51].

Loss of the acute hypocalcaemic response to calcitonin does not prove that the hormone has lost its activity, even where the concentration of anticalcitonin antibodies is high[47]. The hypocalcaemic effect depends on the suppression of osteoclastic function (osteoclastic quiescence), and the long-term effect of calcitonin may be to block the differentiation of osteoprogenitor cells into osteoclasts[46]. Antibody-bound calcitonin might still be capable of exerting its effects on osteoclast proliferation, even in those exceptional cases in which there is a high concentration of antibodies (F. R. Singer, personal communication 1983).

However, in patients, who have become clinically resistant, antibody titres should be determined so as to confirm or eliminate this as the cause.

Finally, in those patients who fail to show a hypocalcaemic response despite a low antibody titre, the reason may be that calcitonin has an exceptional affinity for the antibodies, exceeding its affinity for its specific receptors[38–40,52]. This might be an early indication that antibody-mediated resistance is developing[39] and is attenuating the hypocalcaemic response.

● Inhibition of hypocalcaemia in the rat bioassay system[24,44]

When a calcitonin was incubated with serum containing specific anticalcitonin antibodies and then injected subcutaneously into rats, the hypocalcaemic response was diminished[36–38,40]. When it was injected intravenously, however, the response was not diminished; in fact it was even prolonged[53], suggesting that, like antibodies to insulin in diabetic patients, calcitonin antibodies might even potentiate the hormone's action in some way. Therefore, this approach cannot be considered a valid way of studying the relation between calcitonin, antibodies and hypocalcaemic response.

● Restoration of control over bone turnover

The best way of assessing antibody involvement in the event of resistance is to determine whether or not control over bone turnover can be re-established by giving a type of calcitonin which does not cross-react with the antibodies already present (see Fig. 113). The first step, however, is to ensure that the loss of control is not due to non-compliance on the part of the patient, because both serum alkaline phosphatase and urinary hydroxyproline levels are known to increase rapidly after calcitonin withdrawal, and these changes cannot be distinguished from those observed during the development of resistance[25,45,54].

Most patients respond rapidly when switched to a different type of calcitonin, but in those with very active bone disease the spontaneous fluctuations in bone turnover mean that several months may elapse before it is apparent whether or not control has been restored. A change in the type of calcitonin will only return bone turnover to the level existing before the development of resistance. Where turnover remains high (plateau effect) a change of species is unlikely to reduce it.

Resistance cannot be assessed clinically. Pain, for example, may be due to a variety of factors and shows little correlation with changes in bone turnover. It is more likely to be caused by osteoarthritis or the extension of a fissure fracture than by the development of antibody-mediated resistance[24,55].

Summary: Antibody formation

The relationship between the effectiveness of calcitonin and the presence of calcitonin antibodies may be summarized as follows:

– Antibodies may develop in patients treated with calcitonins other than HCT.

– There is a weak correlation between the dose and duration of treatment on the one hand, and the development of antibodies and their effect on therapeutic response on the other[42,43,56]. This should be borne in mind so that, if necessary, proper measures can be taken to determine whether antibody-mediated resistance really has developed and, if it has, appropriate action taken.

– The risk of resistance developing through the formation of antibodies may be reduced by using an intermittent treatment regimen. In many cases this will minimize the difficulty[24] while maintaining good control over bone turnover. Against this, some authors claim that intermittent treatment may boost antibody titres, as it appears to do in animals.

– Because antibody titre determinations are not reliable, the incidence of genuine antibody-mediated resistance has probably been greatly overestimated. No really objective proof has yet been adduced that calcitonin has greater affinity for its antibodies than for the specific receptors through which it exerts its biological effect. Calcitonin has not been shown to be irreversibly bound to antibody, nor has the calcitonin-antibody complex been shown to be inactive.

– In the present state of knowledge, a change to a type of calcitonin that does not cross-react with the circulating antibody is still the best way of assessing antibody-mediated resistance.

Interactions and physicochemical incompatibilities with other drugs

Parathyroid hormone raises blood calcium, whereas calcitonin lowers it; but both hormones lower blood phosphate levels, and in this respect their actions are additive, so that they increase urinary phosphate levels[57,58]. Apart from this hormonal interaction, a survey of the literature yields the following interactions and incompatibilities involving calcitonin:

– Thyroxine
Thyroxine raises the blood phosphate level, thus abolishing the hypophosphataemia induced by calcitonin. It also counteracts the hypocalcaemic effect of calcitonin by increasing bone resorption[58,59].

– Heparin
Long-term treatment with heparin may give rise to osteoporosis; this can be prevented or reversed by calcitonin.

– Glucocorticoids
Glucocorticoids also tend to cause osteoporosis, which can be prevented or reversed by calcitonin. In addition, combination of a glucocorticoid (prednisone) with calcitonin will reduce, reverse or even prevent the development of resistance (escape phenomenon)[27].

– Pizotifen
This antiserotoninergic drug has been reported to reduce or prevent the side effects of calcitonin[60].

– Theophylline and isoprenaline
These compounds have been reported to inhibit the hypocalcaemic effect of calcitonin in rats[59], although theophylline, an intracellular inhibitor of cAMP degradation which by itself has no effect on osteoclastic motility *in vitro*, potentiates calcitonin-induced osteoclastic quiescence[61].

– Imidazole
Imidazole increases cAMP degradation, thus reducing osteoclast sensitivity to calcitonin.

The last three interactions are all due to the involvement of cyclic AMP as second messenger mediating the action of calcitonin[57].

The following chemical and physical facts should also be noted in handling calcitonin:

– The 1–7 disulphide bridge is reported to undergo oxidation in certain media, methionine in porcine and human calcitonin oxidizing particularly readily.

– Oxidation of the disulphide bridge is also thought to occur in alkaline media.

– At low concentrations calcitonin may be adsorbed onto the walls of its container[57,62]. This phenomenon can be minimized by using a protein carrier and diluting with weak acid[62]. Dimers may be formed as a result of changes (especially falls) in pH.

References

1 Fillastre JP et al: Furosemide, mithramycin and salmon calcitonin in hypercalcemia. Europ J intensive Care Med 1975, 1, 185–8.

2 Hosking DJ: Treatment of severe hypercalcaemia with calcitonin. Metab Bone Dis rel Res 1980, 2, 207–12.

3 Singer FR et al: Paget's disease of bone. In: Metabolic bone disease II. Ed Avioli LV, Krane SM. Academic Press 1978, 489–575.

4 Agrawal R et al: Calcitonin treatment of osteoporosis. In: Calcitonin 1980, Proc Int Symp Milan 1980. Ed Pecile A. Excerpta Med Int Congr Ser 1981, 540, 237–46.

5 Chesnut III CH et al: Calcitonin and postmenopausal osteoporosis. In: Calcitonin 1980, Proc Int Symp Milan 1980. Ed Pecile A. Excerpta Med Int Congr Ser 1981, 540, 247–55.

6 Vigo P et al: Recent progress in the treatment of postmenopausal and senile osteoporosis. In: Calcitonin 1980, Proc Int Symp Milan 1980. Ed Pecile A. Excerpta Med Int Congr Ser 1981, 540, 256–68.

7 Wallach S et al: Effect of salmon calcitonin on skeletal mass in osteoporosis. Curr ther Res 1977, 22, 556–72.

8 Gennari C: Calcitonin and bone metastases of cancer. In: Calcitonin 1980, Proc int Symp, Milan 1980. Ed Pecile A, Excerpta Medica 1981, Int Cong Ser 540, 277–87.

9 Goebell H, Hotz J: Calcitonin, pancreatic secretion and pancreatitis. In: Calcitonin 1980, Proc int Symp, Milan 1980. Ed Pecile A, Excerpta Medica 1981, Int Cong Ser 540, 346–51.

10 Castells S et al: Therapy of osteogenesis imperfecta with synthetic salmon calcitonin. J Pediat 1979, 95, 807–11.

11 Rosenberg E, Lang R, Boisseau V, Rojanasathit S, Avioli LV: Effect of long-term calcitonin therapy on the clinical course of osteogenesis imperfecta. J Clin Endocrinol Metab 1977, 44, 346–55.

12 Hahn TJ, Boisseau VC, Avioli LV: Effect of chronic corticosteroid administration on diaphyseal and metaphyseal bone mass. J Clin Endocrinol Metab 1974, 39, 274–82.

13 Avioli LV, Gennari C: Calcitonin therapy for bone disease and hypercalcemia. In: The effects of calcitonin in man, Proc 1st Int Workshop, Florence 1982, Ed Gennari C, Segre G. Masson 1983, 103–10.

14 Wener JA, Gorton SJ, Raisz LG: Escape from inhibition or resorption in cultures of fetal bone treated with calcitonin and parathyroid hormone. Endocrinology 1972, 90, 752–9.

15 Tashjian AH Jr, Wright DR, Ivey JL, Pont A: Calcitonin binding sites in bone: relationships to biological response and "escape". Recent Prog Horm Res 1978, 34, 285–334.

16 Salmon DM et al: Decreased responsiveness to chronic salmon calcitonin treatment in rat kidney and calvaria studied using quantitative enzyme cytochemistry. Acta Endocrinol 1985, 108, 570–6.

17 Mazzuoli GF et al: Il fenomeno escape e plateau. In: The effects of calcitonin in man, Proc 1st int Wkshp, Florence 1982. Ed Gennari C, Segre G, Masson 1983, 75–84.

18 Hamilton CR Jr: Effects of synthetic salmon calcitonin in patients with Paget's disease of bone. Am J Med 1974, 56, 315–22.

19 Evans IMA: Human calcitonin in the treatment of Paget's disease: long-term trials. In: Human calcitonin and Paget's disease, Proc int Workshop, London 1976. Ed MacIntyre I. Huber 1977, 111–23.

20 Nagant de Deuxchaisnes C et al: Roentgenologic evaluation of the efficacy of calcitonin in Paget's disease of bone. In: Molecular Endocrinology. Ed MacIntyre I, Szelke M. Elsevier 1977, 213–33.

21 Nagant de Deuxchaisnes C et al: Roentgenologic evaluation of the action of the diphosphonate EHDP and of combined therapy (EHDP and Calcitonin) in Paget's disease of bone. In: Molecular Endocrinology. Ed MacIntyre I, Szelke M. Elsevier 1979, 405–33.

22 Nagant de Deuxchaisnes C et al: Comparative effects of calcitonin, the diphosphonate EHDP, and combined therapy calcitonin-EHDP on the radiographic lesions of Paget's disease. Acta rhumatol 1981, 4, 425–57.

23 Nagant de Deuxchaisnes C et al: The action of the main therapeutic regimes on Paget's disease of bone, with a note on the effect of vitamin D deficiency. Arthritis Rheum 1980, 23, 1215–34.

24 Hosking DJ: Practical implications of calcitonin antigenicity. In: The effects of calcitonin in man, Proc 1st Int Workshop, Florence 1982, Ed Gennari C, Segre G. Masson 1983, 67–74.

25 Evans IMA et al: Paget's disease of bone – the effect of stopping long-term human calcitonin and recommendation for future treatment. Metab Bone Dis rel Res 1980, 2, 87–92.

26 De Rose J et al: Response of Paget's disease to porcine and salmon calcitonins: effects of long-term treatment. Am J Med 1974, 56, 858–66.

27 Gennari C et al: Ruolo della calcitonina nell'ipercalcemia acuta maligna. In: The effects of calcitonin in man, Proc 1st int Wkshp, Florence 1982. Ed Gennari C, Segre G, Masson 1983, 187–201.

28 Segre G et al: Non-traditional activities of calcitonin. In: The effects of calcitonin in man, Proc 1st int Wkshp, Florence 1982. Ed Gennari C, Segre G, Masson 1983, 55–64.

29 Cooper CW, Obie JF: Use of the Alzet minipump to infuse calcitonin (CT) and parathyroid hormone (PTH) continuously in the rat. Pharmacologist, 1978, 20, 37.

30 Raisz LG, Au WY, Friedman J, Niemann I: Thyrocalcitonin and bone resorption. Studies employing a tissue culture bioassay. Am J Med 1967, 43, 684–90.

31 Krieger NS, Feldman RS, Tashjian AH Jr: Parathyroid hormone and calcitonin interactions in bone: irradiation-induced inhibition of escape in vitro. Calcif Tissue Int 1982, 34, 197–203.

32 Nakamura T et al: Whole-body irradiation inhibits the escape phenomenon of osteoclasts in bones of calcitonin-treated rats. Calcif Tissue Int 1985, 37, 42–5.

33 Krieger NS, Tashjian AH Jr: Inhibition by ouabain of parathyroid-hormone-stimulated bone resorption. J Pharmacol Exp Ther 1981, 217, 586–91.

34   Feldman RS, Krieger NS, Tashjian AH Jr: Effects of para-thyroid hormone and calcitonin on osteoclast formation in vitro. Endocrinology 1980, 107, 1137–43.

35   Austin LA, Heath H III: Calcitonin: physiology and patho-physiology. N Engl J Med 1981, 304, 269–78.

36   Plehwe WE, Hudson J, Clifton-Bligh P, Posen S: Porcine cal-citonin in the treatment of Paget's disease of bone: experience with 32 patients. Med J Aust 1977, 64, 577–81.

37   Martin TJ, Jerums G, Melick RA, Xipell JM, Arnott R: Clini-cal, biochemical and histological observations on the effect of porcine calcitonin in Paget's disease of bone. Aust NZ J Med 1977, 7, 36–43.

38   Singer FR, Aldred JP, Neer RM, Krane SM, Potts JT Jr, Bloch KJ: An evaluation of antibodies and clinical resistance to salmon calcitonin. J Clin Invest 1972, 51, 2331–8.

39   Woodhouse NJ, Mohamedally SM, Saed-Nejad F, Martin TJ: Development and significance of antibodies to salmon calcito-nin in patients with Paget's disease on long-term treatment. Br Med J 1977, 2, 927–9.

40   Haddad JG Jr, Caldwell JG: Calcitonin resistance: clinical and immunologic studies in subjects with Paget's disease of bone treated with porcine and salmon calcitonins. J Clin In-vest 1972, 51, 3133–41.

41   Singer FR et al: Studies of the treatment and aetiology of Paget's disease of bone. In: Human calcitonin and Paget's dis-ease, Proc int Workshop, London 1976. Ed MacIntyre I. Huber 1977, 93–110.

42   Singer FR, Ahrne-Collier I: Salmon calcitonin therapy: anti-bodies and clinical resistance in patients with Paget's disease of bone. In: Molecular Endocrinology. Ed MacIntyre I, Szelke M. Elsevier 1977, 207–12.

43   Dietrich FM, Fischer JA, Bijvoet OL: Formation of anti-bodies to synthetic human calcitonin during treatment of Paget's disease. Acta Endocrinol (Copenh) 1979, 92, 468–76.

44   Hosking DJ: Clinical significance of calcitonin antibodies. In: Calcitonin 1980, Proc int Symp, Milan 1980. Ed Pecile A, Ex-cerpta Medica 1981, Int Cong Ser 540, 365–77.

45   Avramides A, Flores A, DeRose J, Wallach S: Paget's disease of the bone: observations after cessation of long-term syn-thetic salmon calcitonin treatment. J Clin Endocrinol Metab 1976, 42, 459–63.

46   Singer FR, Fredericks RS, Minkin C: Salmon calcitonin therapy for Paget's disease of bone. The problem of acquired clinical resistance. Arthritis Rheum 1980, 23, 1148–54.

47   Martin TJ: Treatment of Paget's disease with the calcitonins. Aust NZ J Med 1979, 9, 36–43.

48   Bijvoet OL, Sluys-Veer J van der, Jansen AP: Effects of cal-citonin on patients with Paget's disease, thyrotoxicosis, or hypercalcaemia. Lancet 1968, 1, 876–81.

49   Hosking DJ, Denton LB, Cadge B, Martin TJ: Functional sig-nificance of antibody formation after long-term salmon cal-citonin therapy. Clin Endocrinol (Oxf) 1979, 10, 243–52.

50   Guttmann S et al: Distribution of calcitonins between their re-ceptors and antibodies. In: The effects of calcitonin in man,

Proc 1st int Wkshp, Florence 1982. Ed Gennari C, Segre G, Masson 1983, 25–31.

51   Berson A et al: Insulin-I131 metabolism in human subjects: Demonstration of insulin-binding globulin in the circulation of insulin treated subjects. J Clin Invest 1956, 35, 170–90.

52   Rojanasathit S, Rosenberg E, Haddad JG Jr: Paget's bone dis-ease: response to human calcitonin in patients resistant to sal-mon calcitonin. Lancet 1974, 2, 1412–5.

53   Aldred JP et al: Retention of biological activity of antibody bound salmon calcitonin. Fed Proc 1978, 37, 272.

54   Hosking DJ, Bijvoet OLM: Therapeutic uses of calcitonin. In: Endocrinology of calcium metabolism, therapeutic uses of cal-citonin. Ed Parsons JA. Raven Press 1982.

55   Redden JF, Dixon J, Vennart W, Hosking DJ: Management of fissure fractures in Paget's disease. Int Orthop 1981, 5, 103–6.

56   Avramides A, Flores A, De Rose J, Wallach S: Treatment of Paget's disease of bone with once a week injections of salmon calcitonin. Br Med J 1975, 3, 632.

57   Milhaud G: Calcitonine. In: Pharmacologie clinique, Vol 1. Ed Giroud JP et al. Expansion scientifique française, Paris 1978, 859–71.

58   Milhaud G, Moukhtar MS: Antagonistic and synergistic ac-tions of thyrocalcitonin and parathyroid hormone on the levels of calcium and phosphate in the rat. Nature 1966, 211, 1186–7.

59   Wells H, Lloyd W: Inhibition of the hypocalcemic action of thyrocalcitonin by theophylline and isoproterenol. Endo-crinology 1968, 82, 468–74.

60   Gennari C et al: Gli effetti collaterali di differenti calcitonine. In: The effects of calcitonin in man, Proc 1st int Wkshp, Flor-ence 1982. Ed Gennari C, Segre G, Masson 1983, 93–101.

61   Chambers TJ, Azria M: The effect of calcitonin on the osteo-clast. Triangle, Sandoz Journal of Medical Science 1988, 27, 53–60.

62   Parsons JA: Effects of added protein on apparent potency of thyrocalcitonin. In: Calcitonin, Proc Symp on Thyrocalcitonin and C cells. Ed: Taylor S. Heinemann 1968, 36–41.

63   Reginster JY, Albert A, Lecart MP, Lambelin P, Denis D, Deroisy R, Fontaine MA, Franchimont P: 1-year controlled randomised trial of prevention of early postmenopausal bone loss by intranasal calcitonin. Lancet 1987, 2, 1481–3.

64   Christiansen C, Riis BJ, Leuphant J, Johansen SS: Intranasal calcitonin in the treatment of postmenopausal osteoporosis. Osteoporosis 1987, 2, 853–6.

65   Minaro P, Mallet E, Levernieux J, Schoutens A, Attali G, Caulir F: Immobilization bone loss: Preventive effect of cal-citonin in several clinical models. Osteoporosis 1987, 1, 603–10.

Chapter 5: General Conclusions

Since the discovery of calcitonin by Copp in 1961, the hormone has been identified in several animal species, including man. The various types have been isolated and their structures elucidated, and some of them, in one case an analogue, have been synthesized. Four types are currently in therapeutic use, the two derived from fish being considerably more potent than the mammalian varieties, human and porcine calcitonin.

All this has generated a vast body of literature on calcitonin which, however, contains many discrepant findings regarding both the physiological role and the pharmacological properties of the hormone. Although many of these inconsistencies are probably more apparent than real, it is nevertheless prudent to continue to describe calcitonin conservatively as a hormone which, in combination with many other factors, has a 'corrective and/or regulatory' effect on the following processes:

- Phosphorus and calcium metabolism under normal physiological conditions – for example the maintenance of calcium balance and skeletal mass. Calcitonin does this by controlling bone remodelling and the uptake, storage and elimination of calcium. It is also active in situations involving high calcium demand such as occur during bone growth, pregnancy and lactation, and eating.

- Gastric and pancreatic function

- Renal function, especially excretion of water and electrolytes

- Certain endocrine functions, such as prolactin secretion

- Certain CNS mechanisms (pain control and its neuromodulatory role)

- All biological processes in which the movement of calcium has a key role

In therapeutic use the calcitonins have demonstrated a low level of toxicity, with relatively minor side effects such as facial flushing and nausea. They therefore offer a wide margin of safety. Their efficacy in diseases characterized by disorders of bone remodelling or bone loss such as Paget's disease, some types of hypercalcaemia, high-bone-turnover osteoporosis and pain associated with osteoporosis is well established, and they will very probably also prove useful in other diseases involving bone such as periodontosis, as well as in non-skeletal diseases, especially of the cardiovascular system (immunoarteriosclerosis, migraine, arrhythmias) and the nervous system.

A major prerequisite for such extension of the use and usefulness of calcitonin-based drugs, however, is the improvement of patient acceptability. Especially where long-term treatment is involved – as it inevitably is in many of these conditions – an oral or other non-injectable dosage form is absolutely essential if calcitonin is to realize its full therapeutic potential. The recent development of nasal-spray and suppository forms are a step in this direction, and clinical trials indicate that these new dosage forms are likely to offer an effective alternative to present forms of the drug.

Subject Index